FOOD AND DEVELOPMENT IN THE SEMI-ARID ZONE OF EAST AFRICA

by

Philip W. Porter

Foreign and Comparative Studies/African Series XXXII

Maxwell School of Citizenship and Public Affairs
Syracuse University
1979

Library of Congress Cataloging in Publication Data

Porter, Philip Wayland, 1928-
Food and development in the semi-arid zone of
east Africa.

(Foreign and comparative studies : African series ;
32)
Bibliography: p.
1. Arid regions agriculture--Africa, East.
2. Food supply--Africa, East. 3. Diet--Africa, East.
4. Nutrition policy--Africa, East. I. Title.
II. Series
S616.A44P67 630'.9676 79-20312
ISBN 0-915984-54-7

CONTENTS

LIST OF TABLES

LIST OF FIGURES

LIST OF PLATES

INTRODUCTION[1]

> On leaving Kaptè we entered upon the mountainous district of Ulu, which we found to be densely inhabited, fertile, and well cultivated, with cattle also in great numbers. (Plate 1) In a few hard marches I traversed this friendly district, with rising hopes of life, and dreams of home and friends. Instead of becoming worse, I was getting better, and the only bar to a rapid recovery from my state of emaciation was the absence of any digestible food.
>
> Leaving Ulu, we emerged on the barren wastes [Makueni] which stretch away to Kikumbuliu, and at

---

[1]Some of the ideas developed herein were presented in July 1977 at a workshop on Drought Preparedness sponsored by the National Academy of Engineering, preparatory to the U.N. Conference on Desertification. I thank the following individuals for bibliographic assistance as well as the use of some of their research findings: Carl Eicher, Department of Agricultural Economics, Michigan State University; James Kocher, Institute Associate at the Harvard Institute for International Development; James Newman, Department of Geography, Syracuse University; Ben Wisner, Universidade Eduardo Mondlane, Maputo, Mozambique; and Edgar V. Winans, Social Science Advisor, Ford Foundation, Nairobi, Kenya.

Many books have been written in recent years about hunger, nutrition, food and agriculture. Two kinds of references are listed: general works on agriculture, food and nutrition, and similar works dealing with East Africa. Full citations are given in the bibliography: (1) general works on agriculture, food and nutrition, written from a variety of points of view: Abelson, 1975; Agency for International Development, 1976; Aylward and Jul, 1975; Berg, Scrimshaw and Call, 1973; DeMarco and Sechler, 1975; deWilde, 1967; Knight and Wilcox, 1976; Latham, 1965; Rechcígl, 1975; Scientific American, 1976; Scrimshaw and Béhar, 1974; Southworth and Johnston, 1967; White, 1974; Willett, 1976; and World Food Conference, 1977; (2) books and articles on East Africa which contain useful bibliographies on agriculture, food and nutrition: Amann, Belshaw and Stanfield, 1972; Bohdal, Gibbs and Simmons, 1968; Kocher, 1977; Kraut and Cremer, 1969; Latham, 1967; Lele, 1975; Mascarenhas, 1973; May and McLellan, 1970; McLoughlin, 1967; and Ruthenberg, 1968.

> a killing rate we rushed through this forbidding, uninhabited wilderness--for our goods were exhausted, and the men were on half rations....
>
> We were once more in the 'Nyika,' with all its inevitable horrors. (Plate 2) We crossed Kikumbuliu, and found the people dying of famine; so no food was to be got there (Thomson, 1885: 573-575).[1]

Joseph Thomson was traversing Machakos District, Kenya, going from well-watered land to semi-arid country, an area still plagued with persistent agricultural difficulties. This paper about East Africa was originally written to "stimulate thought on world food policy in terms of the necessity to account for particular geographical circumstances, the unique societal conditions associated with geography and policy itself" (R. H. Myers, 1976).[2] More particularly the paper was to focus on state policy regarding famine arising in areas of rainfed agriculture that suffer severe climatic variation. Although this essay will focus on the livelihoods of people living in semi-arid areas of East Africa, it will try to look at these areas in a number of broader contexts: (1) with regard to the people and economies of the wetter environments at their margins; (2) with regard to the nations and national policies within which these areas exist, and (3) with regard to the political and technological economy of the world.

---

[1]The famine encountered by Thomson may have been the Yua ya Ndata ("famine of the star") which Munro (1975: 23) links to the Great September Comet of 1882. Famine conditions may still have prevailed in May 1884, when Thomson passed through the Machakos-Kikumbulyu area.

[2]This study was originally written under commission from the Hoover Institution to form one of three papers in a volume to be titled Food Production Crises in Dry Farming Regions of the World. Since one of the authors did not complete a manuscript the project was abandoned.

The production of food through farming, herding, fishing, and collecting, occupies the energies and time of the vast majority of East Africa's people, perhaps as much as 90 percent of the population. Of course, people spend much time in other productive and not-so-productive ways--building and repairing houses, teaching children, sewing, carrying water, talking, attending meetings, going to market, and sharing beer with neighbors. (Plate 3) To be a successful farmer or herder means more than to feed a family and provide for its other needs. Success may confer high status and respect, and ensure control over resources (particularly good land and the labor of others). In the case of a man it may enable him to take a second or third wife, to become politically prominent, and to expand into other fields, such as transport and commerce. Further, arts and aesthetics are often found to be integrally bound up in the business of living and are inseparable from productive work. They can be part of a cosmography which does not distinguish between sacred and secular, and indeed may not divide the world into separate domains--man versus nature--as is done in Western societies. A change in the manner in which food is produced reverberates throughout the system, affecting relations of people with one another and with other aspects of the culture. When we study "the unique societal conditions" of food production and policy in East Africa, it becomes clear that our approach needs also to be historical.

We approach the topic of food supply in East Africa through a sequence of scales of increasingly finer resolution descending from the global to the child. The food supply question is usually posed in terms of what "we" who live in developed Western industrial countries could do for "them"--people of underdeveloped countries--by sharing agricultural technology so that they too might enjoy a plentiful, reliable food supply. For the family of Kitemu wa Nguli (to whom we return) in the semi-arid zone of Kenya on the margins of the Nyika near Kikumbulyu, the adequacy of the food supply is not a matter for abstract scholarly speculation, econometric modelling, and global forecasting. (Plate 4) It is life itself--a continuously unfolding, unremitting event, in which the state of food supply or its prospects may change from day to day. The food status of Kitemu and other people living in East Africa's semi-arid areas in 1984 is likely to be no better than that encountered by Joseph Thomson in May 1884 in Kikumbulyu. This essay will explore the complex set of reasons why after a hundred years this should still be so, and will ask how it might be different.

Global and National Perspectives

Although international grain trade does not measure national self-sufficiency in food production, or the dietary status of individual citizens, it does point to global trends in food dependency (Table 1).

For Africa as a whole there was equilibrium in grain trade in the 1930s and 1940s, but in the 1960s and 1970s the situation became one of increasing food imports. A continental statistic masks great variation within countries and includes the grain

TABLE 1

WORLD GRAIN, 1934-76 SELECTED PERIODS

(in millions of metric tons;
+ export, - import)

Region	1934-38	1948-52	1960	1970	1976 (est.)
North America	+ 5	+23	+39	+56	+94
Latin America	+ 9	+ 1	--	+ 4	- 3
Western Europe	-24	-22	-25	-30	-17
Eastern Europe and USSR	+ 5	--	--	--	-27
AFRICA	+ 1	--	- 2	- 5	-10
Asia	+ 2	- 6	-17	-37	-47
Australia and New Zealand	- 3	+ 3	+ 6	+12	+ 8

Source: Burki and Goering, See also West, 1974.

surpluses of South Africa, with deficit areas in west, central and eastern Africa. We move next to the country scale.

East Africa is generally taken to be Kenya, Tanzania, and Uganda. At a national level, all three East African countries had a balanced grain trade throughout the 1950s and during much of the 1960s (Figure 1). After 1968 each country exhibited a different pattern. Kenya developed a positive balance, exporting major amounts of maize, rice and wheat in most years. Uganda developed a chronic negative balance which cost an average $2.75 million in imports each year for the period 1968-1975. Tanzania had a small negative trade balance in grains, but in 1972 a series of bad years commenced costing the Tanzanian people in four successive years (1972-1975) 3.0, 0.9, 22.0 and 34.7 million dollars (U.S.) in foreign currency, a significant percentage of its total net foreign exchange earnings. From outward appearances, it would seem that Kenya generally has no food shortage it cannot solve by local distribution of grain it normally exports, but that Uganda and especially Tanzania have developed serious food shortages in the past decade. Outward appearances in this instance deceive us,[1] we must move to a still finer scale, and look particularly at food production in the semi-arid zone of these countries.

The Semi-Arid Zone and Its Population

Any definition of the semi-arid area of East Africa is arbitrary. For this analysis I have chosen a definition that is

---

[1]At times a significant part of the maize commercially grown in the former European highlands has been exported to Europe to be made into laundry starch and used as fillers and extenders in ice cream and other products that use cornstarch. Exports of this kind occurred in the early 1970s, contemporaneously with food shortages and imports caused by drought in northern and eastern Kenya. Since 1977 export of maize has required the permission of the Minister of Agriculture. In the resulting uncertainty as to whether a coming season's maize crop could be exported large scale farmers (and even some small scale farmers) in the former European highlands have planted less maize and shifted to uncontrolled crops, such as sunflowers and other oil seeds (Muthoka, 1978: 5).

FIGURE 1

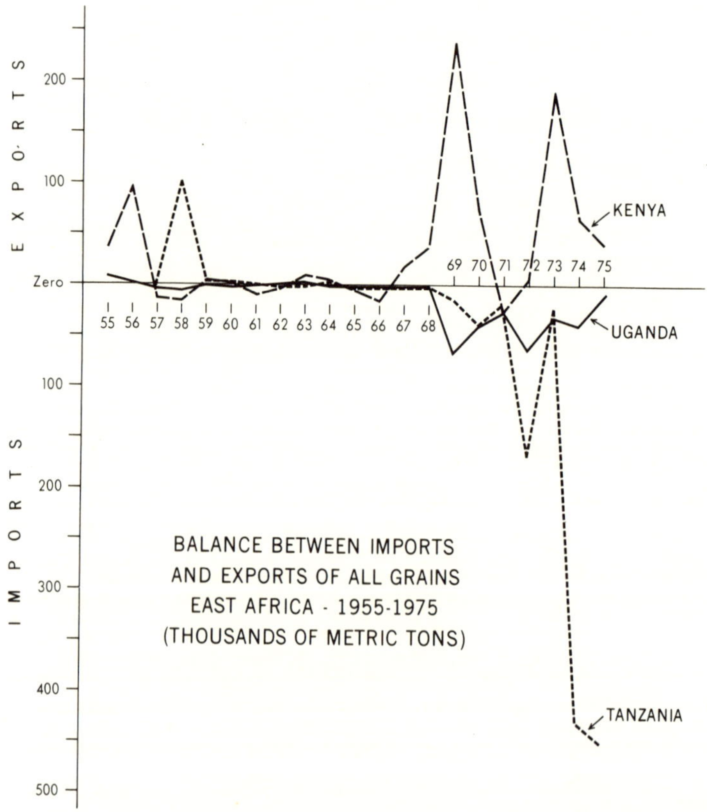

Source: UN FAO Trade Yearbook - 1959, 1963, 1969, 1974 and 1975.

easily remembered and which clearly separates those people whose agriculture is at special risk of climatic variation from those in better watered areas and from those in drier areas where rainfed agriculture is not possible and where pastoralism is the main livelihood. The wet and dry margins of the zone are defined respectively as 800 and 400 mm (32 and 16 inches) of rainfall (annual mean).[1]

Rainfall, of course, is one variable embedded in a set of other variables, which all together determine whether or not a plant will bear its seed. Our simple definition leaves unanswered questions of whether rain is concentrated in one season or divided into two, of the moisture demand (potential crop transpiration) at the place, of soil character, of crop phenology, and omits a host of questions having to do with crop and field management--interplanting, shading, timing, and the like. The variability of observed rainfall around monthly and annual means emphasizes the fact that we are dealing with a gradient. The nearby humid environments of East Africa are different in degree, not in kind. They are not free of the climatic variability which is so important to farmers in the sub-humid and semi-arid margins. As we will see, the 400-800 mm rainfall zone is "marginal" in several ways, not only environmentally, but economically, socially and politically.

Figure 2 shows the 800 and 400 mm rainfall isolines superimposed on a dot map of East Africa's population as of 1962. This first look shows us that the semi-arid zone is found almost wholly in Kenya and Tanzania. In Uganda small fragments of semi-arid country occur in the northeast in Karamoja, in the rift valleys of Lakes Mobutu Sese Seko (Lake Albert) and Idi Amin (Lake Edward),

---

[1]A more precise way to use these terms would be to compare mean annual rainfall with potential evaporation (E_o, Penman's potential open water evaporation) at each place to decide on whether a place on average is humid, sub-humid, semi-arid or arid (Penman, 1963). Since periods significant for crops in the semi-arid areas range from about two to five months in this zone, comparison of annual rainfall and E_o values would tell us little more than our definition based on rainfall. The semi-arid zone has been variously defined (Ominde, 1971; International Labour Office, 1972; Woodhead, 1968, Wisner, 1976; and Hare, 1977).

FIGURE 2

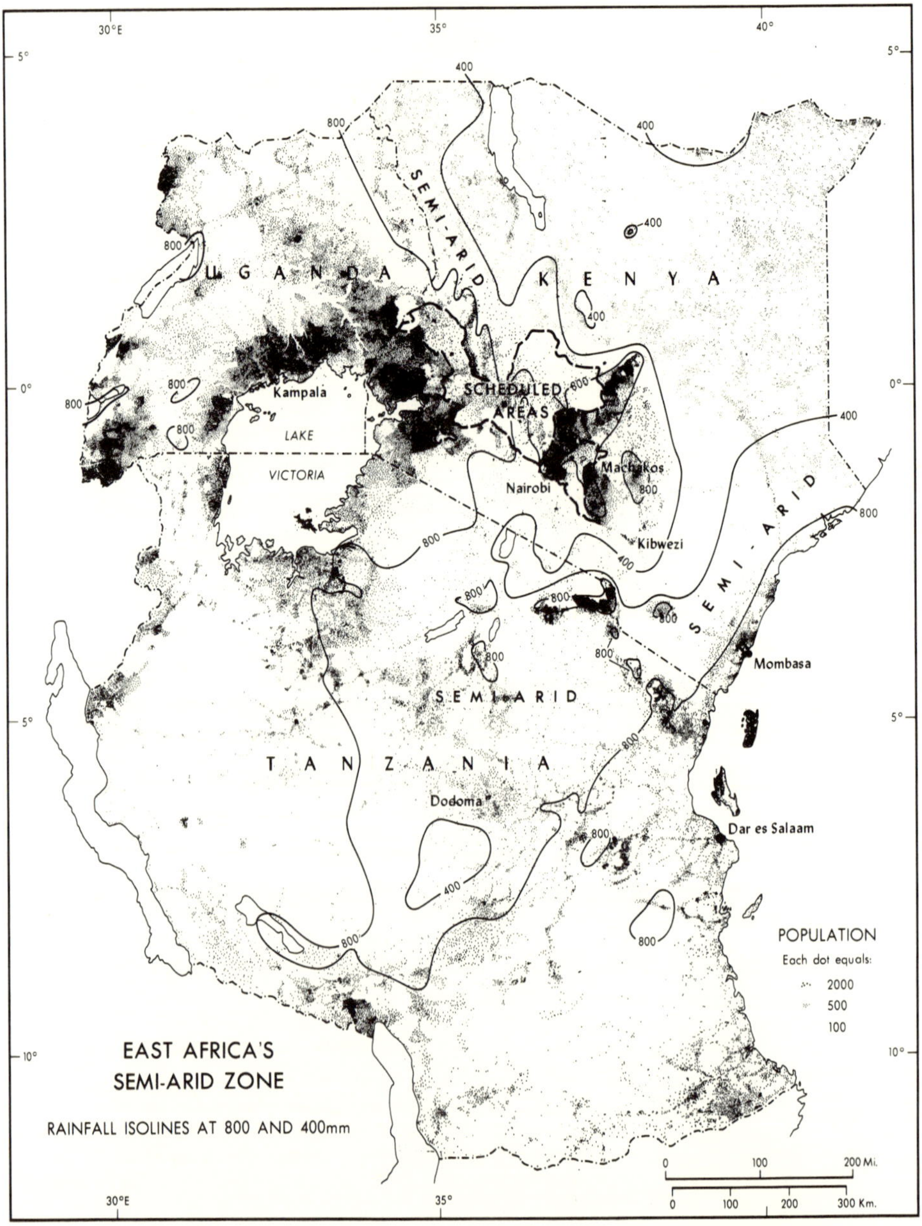
EAST AFRICA'S
SEMI-ARID ZONE
RAINFALL ISOLINES AT 800 AND 400mm
POPULATION
Each dot equals:
2000
500
100
UGANDA
KENYA
TANZANIA
SEMI-ARID
SEMIARID
SCHEDULED AREAS
Kampala
LAKE
VICTORIA
Nairobi
Machakos
Kibwezi
Mombasa
Dodoma
Dar es Salaam
30°E
35°
40°
5°
0°
10°
0
100
200 Mi.
300 Km.

and in southwestern Uganda (eastern Ankole). In Kenya there are two semi-arid regions, one is coastal, merging into "the 'Nyika,' with all its inevitable horrors," and the other is what has been termed the Eastern Foreland Plateau (Wisner, 1976: 5), a zone generally from 500 m to 1,500 m which surrounds the high-potential heartland of Kenya. Its sinuous course occupies Turkana country in the north, crosses West Pokot and Baringo districts, swings through Laikipia, Nanyuki and Meru, takes in most of Kitui and Machakos districts, and then swings through Kajiado and Narok districts with a salient thrusting northwards along the floor of the rift valley around Naivasha. In Tanzania the semi-arid zone comprises the well-known interior wedge of dry country, broad at the Kenya border, tapering to the southwest. It occupies the Masai Steppe, much of the eastern interior plateau (Singida and Dodoma regions), and reaches as far south as the Lake Rukwa Plain and the Bohoro Flats. The wedge is interrupted in a few places by high country where rainfall exceeds 800 mm. There are two small outliers of semi-arid country to the southeast, Mikumi Game Park and in the northern part of the Selous Game Park. The most striking difference between the distribution of semi-arid country in Kenya and Tanzania is that in Kenya the semi-arid zone lies <u>between</u> arid and wet areas and the transition between semi-arid and moist areas is abrupt, often accomplished over just a few miles. The same rapid transition occurs in places in Tanzania--in the northeast where a line of mountainous country includes Mt. Meru, Mt. Kilimanjaro, the Pares, and the Usambaras, and on the northwest margin of the Southern Highlands. Elsewhere in Tanzania, particularly along the western border of the wedge, the gradation into wetter country is gradual. Steepness and shallowness of gradient have consequences for human livelihood because of variations in the proximity of differing kinds of environment to one another. Some East African societies take advantage of differing environmental possibilities where steep gradients, related to relief, place them near one another (Porter, 1965).

How many people live in the semi-arid zone and what sort of demographic future can we predict for the area? Analysis of several census volumes permits us to show how population in the semi-arid zone has increased in the post-independence period, and it provides a basis for population projection. Table 2, based on census data for areas lying within the 400 to 800 mm rainfall zone, tells us how many people are exposed to special kinds of risk in farming the areas where they live.[1]

In 1962, at about the time Kenya, Tanganyika, Uganda and Zanzibar became independent, the population of east Africa stood at 25.8 million. By 1980 it will be about 48 million, and is expected to exceed 90 million by the end of the century. In 1962 about four million people lived in the semi-arid parts of East Africa, two-thirds of them in Tanzania. The average population density in such areas was nine people/sq. km (23 people/sq. mi.).

---

[1]The population of the semi-arid areas was calculated in the following way. The 800 and 400 mm rainfall isolines were superimposed on a census subdivision map for the three East African countries. Populations of census units wholly within the semi-arid zone were counted in full. Census units partly in the semi-arid zone were prorated to the nearest tenth according to their areas, and by extension their population. (Some allowance was made in a few areas where it was known that proportionally more of the population of a unit lived in higher, wetter areas, Mt. Kilimanjaro being a case in point.) Census dates were: 1957 for Tanzania, 1958 for Zanzibar, 1959 for Uganda, and 1962 for Kenya. The populations were brought to two common dates: 1962 and 1980 by means of continuous compounding ($FP=PP.e^{it}$, where FP is future population, PP is present population, i is annual rate of increase, t is the number of years, and e is the exponential function), using values of net increase per annum for the various countries and for various periods: Kenya: 1962-1980, 3.4%; Tanganyika, 1957-1962, 2.2%; Zanzibar and Pemba, 1958-1962, 1.0 and 1.7%; Tanzania, 1957-1967, 3.04%. 1967-1980, 2.7%; and Uganda, 1959-1962, 2.5%, 1962-1969, 3.9%, and 1969-1980, 3.3%. Population growth in semi-arid areas was assumed to equal the national rate. An independent check of the correctness of this assumption was made for Tanzania, where the populations and areas of the semi-arid zone were compared at two censal dates, 1957 and 1967. There was, despite many census boundary changes between censuses, a 98 percent agreement in the area calculation (252,698 sq. km in 1957 versus 247,292 sq. km in 1967) and a close agreement between a population projected for the semi-arid area in 1967 from 1957 data (3,049,000) and the population counted within the semi-arid area using 1967 census data (3,056,000).

TABLE 2

POPULATION AT RISK - THE POPULATION OF THE SEMI-ARID ZONE OF EAST AFRICA

	Population in 1962 (000)	Population in 1980 (proj.) (000)	Percent land area	Percent pop. at risk (1962)	Pop. density 1962 (pop./sq. km)	Pop. density 1980 (proj.) (pop./sq. km)
Kenya	1,235	2,230	31.3	14.3	7.0	13.1
Tanzania	2,619	4,330	28.5	23.1	10.4	17.1
Uganda	133	280	11.6	2.1	6.5	12.3
East Africa	3,987	6,930	27.4	15.5	8.8	15.3

Sources: *1967 Census of Population*, Tanzania, Vol. 1, 1969; Porter, 1966; P. F. Myers, 1976.

By 1980 there will be nearly seven million people living in this same area, and the population density will have risen to 15 people/sq. km (about 40 people/sq. mi.).

Given the differing agricultural possibilities in an area of 800 mm compared with one of 400 mm, one would expect population densities to vary considerably, and they do. A count was made of all census units (location level) which lay entirely within the 400-800 mm rainfall zone. There were 188 such units in 1962; 114 (61%) had densities below 50 people/square mile (about 20 people/square kilometer), 44 (23%) ranged between 50 and 100 people/square mile (20 to 40 people/square kilometer), and 30 (16%) exceeded 100 people/square mile. The highest rural density noted was 300 people/square mile (120 people/square kilometer).

Since births, deaths and migration shape the population character of the semi-arid zone, it is necessary to try to describe what has happened in recent years and to speculate about the area's probable demographic future. Two regions which occupy the central dry wedge of Tanzania (Dodoma and Singida) serve to represent the East African semi-arid zone with respect to its demographic character (Table 3).

In the semi-arid zone, birth rates, death rates and total fertility rates do not differ greatly from the national average, but infant mortality does, being notably higher in some districts--Manyoni 184, Singida 189, and Mpwapwa 164--and notably lower in others--Kondoa 107 (Thomas, 1972). There is, however, a generally higher rate of regional net outmigration, especially of males aged 20 to 40. At the same time, these are areas of new settlement, with families taking up more marginal low-rainfall land for farms. Further, there is a return migration to these areas, especially of older people, which to some degree offset outmigration. Figure 3 illustrates in- and out-migration for Singida Region.

A major phenomenon of population in East Africa is the tremendous amount of redistribution occurring at all times. For example, in Tanzania the 1967 Census of Population showed that 40 percent of the population lived in a place different from the place of birth, and 92.4 percent of these migrants had moved to a rural

TABLE 3

SELECTED DEMOGRAPHIC FACTS FOR TWO SEMI-ARID REGIONS IN TANZANIA

	Dodoma	Singida	Mainland Tanzania
Crude birth rate (0/000)	61	55	57
Crude death rate (est.) (0/000)	22.9	27.4	24.4
Infant mortality rate (0/000)	136	169	139
Total fertility rate (0/000)	6.9	6.1	6.6
Regional net migration (0.000)*	- 1.7	- 9.0	
Interregional efficiency**	- 0.14 low out	- 0.44 high out	

*Regional net migration is in- and out-migration as a percentage of native born population.

**Interregional efficiency is defined as regional net migration divided by regional gross migration: "in - out"/"in + out" (Claeson, 1974: 109).

Sources: Claeson, 1974; Egero and Henin, 1973; Thomas, 1972.

destination. In other words, "rural to rural" migration is the dominant form, and most of it occurs over short distances (Claeson, 1974; Kocher, 1977). In certain parts of Machakos and Kitui districts, east of Nairobi, population growth, because of in-migration, is up to ten times (33%) the national average (Wisner and Mbithi, 1972: 13). Makueni, Kakueni, and the Yatta Plateau, as well as Kwale, Kilifi, Tana River, and Taita districts toward the coast, have become important destinations as population redistributes itself toward drier areas (Ominde, 1968; Mbithi and Wisner, 1973: 140).

The fact that many adult males migrate elsewhere for work gives age-sex pyramids for semi-arid areas an unusual shape. Representative age-sex distributions in Kenya's semi-arid zone in the 1960s are provided in Table 4, with areas of high male employment and the national average for Kenya included for contrast. The

FIGURE 3

AGE STRUCTURE OF REGIONAL MIGRATION STRATA

SINGIDA REGION, CENTRAL TANZANIA

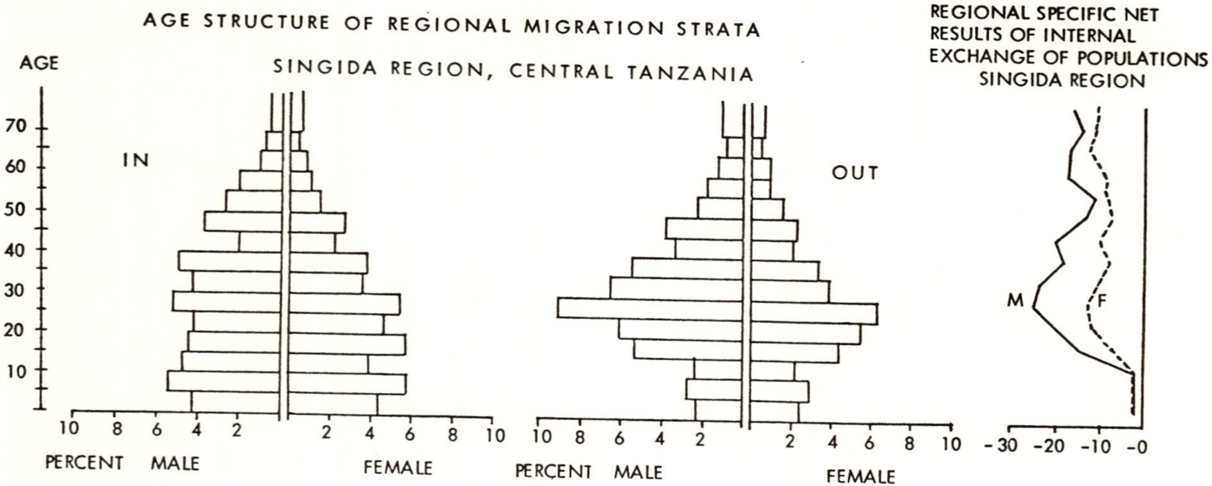

SOURCE: CLAESON, 1974

table speaks for itself. Semi-arid areas where agriculture is risky and food shortages common are the domain of women, children and the elderly. (Plate 5)

TABLE 4

AGE-SEX DISTRIBUTION FOR SELECTED PLACES IN KENYA, 1962

	Population	Adult Males	Adult Females	Children
		(Figures are percentages)		
Sub-humid/semi-arid areas, Machakos District				
Muthetheni Location	16,800	16	29	55
Mukaa Location	27,773	16	27	57
Kalama Location	19,207	15	26	59
Contrasting urban and plantation areas in Kenya Highlands				
Kericho (tea estates)	43,602	48	18	34
Nairobi City	266,794	41	23	36
Nairobi Peri-urban	47,966	48	20	32
All Kenya	8,636,362	23	26	51

Source: <u>Kenya Population Census, 1962</u>, 1964.

The net effect of these various demographic events is that population growth in the semi-arid zone overall keeps pace with, and locally outruns, the national growth of population. The manner in which the increased presence of people in the semi-arid zone has been translated into land use has had profound environmental and social effects. Following a section which describes the bio-physical variability of the semi-arid zone, we will examine the way people have fashioned livelihoods there.

CHAPTER I

THE CLIMATE OF THE SEMI-ARID ZONE AS RESOURCE

Carl Sauer said that a natural resource is a cultural achievement. For Robert Lowie:

> [W]hen every possible allowance is made, the obvious fact still stands that societies have attained solutions of practical problems. Further, this adaptation to the conditions of life constitutes the outstanding intellectual achievement of mankind (Lowie, 1937: 261).

We must look at East African environments in two ways. The first way is to describe its character in the language of western natural science. In so doing, the environment can be understood in comparison with other environments known to the reader. The second way is to describe the character of the environment as seen through the eyes of members of various societies in East Africa. African farmers and herders have, as Lowie suggested, attained solutions to practical problems. Their understanding of environment bears a functional relation to their use of it.

From a western science viewpoint, we may consider East Africa as a laboratory in which agriculture is the experiment to be conducted. In an experiment, one wishes to vary one condition and keep others constant. By moving from place to place in East Africa we can do this, because many of the variables which affect the growth of crops (temperature, radiation, illumination, rainfall) have high degrees of regularity, particularly in their variation with altitude. The following paragraphs attempt to characterize the regularity, variability, and predictability of climatic variables in East Africa, and thus permit us to state the meaning or consequence of each variable.[1]

---

[1]For a somewhat fuller treatment of these variables as they relate to agriculture, see Porter, 1976, which has been drawn upon in writing this section.

Temperature. Temperature, or sensible heat, varies so regularly and directly with altitude in East Africa (the higher, the cooler) that it can be accurately predicted with simple equations.[1] Temperatures at the equator vary little from month to month, although diurnal range is greater with increases in elevation. The greater the latitude, the greater the range in mean temperatures between warmest and coolest month. The agricultural consequences of temperature are discussed later in relation to photosynthesis.

Radiation, Evaporation and Moisture Need. The "hidden twin" of sensible heat is latent heat, or evaporation. Virtually all radiant energy reaching East Africa is transformed in one of two ways; either it does the work of evaporation (it takes 590 gram cal/cm^2 to convert one cc of water from liquid to vapor state), or it heats the atmosphere.[2] In the semi-arid zone, annual potential evaporation (E_0, open water surface evaporation) generally ranges between 1,900 and 2,300 mm annually, the lower values at higher elevations and the higher values at lower elevations.

Space limitations permit only a sketch of the burden radiation places on plants.

> Transpiration of water through leaf pores, or stomata, is the way internal temperatures of plants are regulated. When water is freely available, the rate of transpiration is controlled by the rate of incoming radiation. However, if root and stem systems cannot transport moisture to the leaf at a rate equal to the potential transpiration rate, or if water from the soil becomes a limiting factor, the stomata will

---

[1]For example, the mean maximum temperature is given by the equation T_{max}(°C) = 33.9 - 5.6A; and the equation for the mean minimum temperature is T_{min}(°C) = 24.4 - 6.9A, where A = altitude in thousands of meters.

[2]Evaporation here includes transpiration of moisture from plants. The amount of energy that goes into photosynthesis or is stored as heat in the ground and its cover (plants and animals) is very small, less than two percent.

close, cells will collapse, and the wilting point will be reached. The importance of these facts is that plants themselves do not control their growth or the rate of transpiration of moisture to the atmosphere; they are the passive instruments of transpiration whose rate, when water is nonlimiting, is a direct function of radiation (Porter, 1976: 122-123).

As a rough approximation a field with a crop at full vegetative stage transpires moisture (E_t) at the rate at which moisture evaporates from an open water surface (E_o). Put another way, the ratio $E_t/E_o = 1.0$. Now let us see what these radiation loads mean for some representative places in the semi-arid zone. I have chosen Machakos, Kenya, near the 800 mm margin, Kibwezi - Dwa Plantations, Ltd., in the middle of the semi-arid zone, and Dodoma, in the heart of the semi-arid zone of Tanzania (Table 5).[1]

How, the reader may wonder, can agriculture be successful at any of these places? First, agriculture is only partly successful; and second, it is done by choosing crops that mature quickly, generally have E_t/E_o rates less than 1.0, and are drought tolerant. Some types of bulrush millet, muveta for example, can be grown in 90 to 100 days. Assuming a 3 month season, its E_t/E_o rates might be: 0.6 the first month, 0.9 the second month, and 0.6 the third month. If we apply this crop to two possible seasons in Kibwezi (Table 6), the "grass rains" peaking in November, and the "main rains" peaking in April, we see that on the average there is perhaps nearly enough rainfall to meet crop water needs in the grass rains, but that the main rains on average are insufficient to meet needs. This description does not take into account soil moisture storage, planting population densities, and many other farming practices (to be described later) which reduce the crop's moisture need.

---

[1]Machakos is actually on the wet side of the 800 mm line, in a valley at 1,645 m nestled among the high potential hills of western Machakos District.

TABLE 5

MONTHLY POTENTIAL EVAPORATION AND RAINFALL FOR SELECTED STATIONS IN KENYA AND TANZANIA

	Jan	Feb	Mar	Apr	May	Jun	Jul	Aug	Sep	Oct	Nov	Dec	Total
Machakos													
E_o (mm)	190	174	182	151	140	129	128	140	169	180	158	166	1907
Rainfall (in mm)	50	53	124	207	77	14	6	6	9	53	186	123	893
+ diff	---	---	---	56	---	---	---	---	---	---	28	---	
Kibwezi													
E_o (mm)	170	187	197	176	166	158	156	162	179	191	184	165	2091
Rainfall (in mm)	37	32	92	120	25	5	1	2	4	23	160	140	641
+ diff					none								
Dodoma													
E_o (mm)	167	153	160	149	160	150	160	188	210	229	209	188	2123
Rainfall (in mm)	132	108	119	53	5	1	0	0	0	3	11	104	536
+ diff					none								

Source: Woodhead, 1968a and 1968b.

TABLE 6

BULRUSH MILLET'S WATER REQUIREMENT, KIBWEZI, KENYA

	Oct	Nov	Dec	Jan	Feb	Mar	Apr	May	Jun	Jul	Aug	Sep
		Grass rains--				Main rains--						
E_o in mm	191	184	165	170	187	197	176	166	158	156	162	179
Rainfall in mm	23	160	140	37	32	92	120	25	5	1	2	4
E_t/E_o schedule		0.6	0.9	0.6		0.6	0.9	0.6				
Moisture need ($E_t/E_o \cdot E_o$)		110	149	102		118	158	100				
Difference		+50	- 9	-65		-26	-38	-75				

Radiation and Potential Photosynthesis. Another difference between semi-arid areas and cooler, wetter areas needs to be made, namely the exceedingly important contrast in net photosynthesis (and thus crop yield) between highlands and lowlands. Put simply, for a given crop, highlands equals high yield and lowlands equals low yield. The semi-arid zone is mostly intermediate to low in elevation. Some background is required to explain this contrast in yields.

Photosynthesis is a process in which chlorophyll uses visible light to produce carbohydrate out of water and carbon dioxide. The resulting carbohydrate can be measured as grams/m^2/day (the rate at which an area of plants assimilates the products of photosynthesis), as dry matter production, or as yield of the economically valuable part of the plant. Photosynthesis at a place is a result in part of how much radiation reaches the ground (which is governed by day length, sun angle, cloud cover, aspect [slope exposure], and slope), and the thermal regime; it is also, in part, governed by the characteristics of the crop: its photochemical resistance, resistance in the plant tissue to the diffusion of carbon dioxide, density of foliage, the geometry of the foliage (which governs light interception and light distribution within the canopy), spectral changes in light which has passed through leaves, and the crop's characteristic rates of dark (nighttime) and photo (daytime) respiration. These various features of crops and radiation are known and permit us to describe regional differences in potential photosynthesis (Porter, 1974, 1976; Monteith, 1965, 1972; Wang'ati, 1969).

Looked at in terms of crop yields (Table 7) high elevations have a two to three times advantage over middle and low altitude areas. The causes are partly that potential photosynthesis in the highlands is greater (mainly because cooler nighttime temperatures reduce dark respiration, and thereby more carbohydrate is stored), and partly that the highland growing season is longer, and the number of days a grain crop assimilates in seed is greater. Farmers must choose crops which match the environment crops are grown in. Local Farmers' maize shown in Table 7 has resulted from a process of continual selection of seed to plant next season. The tendency is to select seed which does best in the local area. The hybrid high-altitude maize H611C, which yields 71.1 quintals/hectare in the highlands, takes too long to grow in the drier parts of the semi-arid zone, and is still in a leaf stage when the rains give out. Local Farmers' maize or the quick-growing Katumani, although they give lower yields, complete crop cycle in semi-arid areas quickly and reliably, and thus perform better for farmers in the long run. Nonetheless, H611C grown in the highlands gives double the yield (2.18) of Local Farmers' maize grown in the middle elevations, and it out yields Local Farmers' maize at the coast by a factor of nearly three (2.77).

TABLE 7

OBSERVED MAIZE YIELDS* FOR THREE ALTITUDINAL ZONES IN EAST AFRICA

Maize Type	Over 1650 m (highlands)	900 to 1600 m (intermediate)	0 to 850 m (lowlands)	Yield advantage of Highlands over Intermediate and Lowland zones (Figures are percentages) Intermediate (base 100)	Lowland (base 100)
Local Farmers'	49.9	32.6	25.7	153	194
Katumani & M6	24.4	35.6	22.3	68	109
Ukiriguru Composite A, Cycle 3	61.2	41.2	35.2	149	174
H611C	71.1	45.3	33.8	157	210
H613B	69.6	40.4	29.3	172	238

Source: Porter, 1974. See also Darrah and Penny, 1974.

*Yields in quintals/hectare. A quintal is 100 kg or 220 pounds; a hectare is 2.471 acres.

These contrasts in potential photosynthesis are built-in environmental differences which no amount of plant breeding, soil engineering, or changes in agricultural practice can undo. Just as the mid-latitudes have a built-in yield advantage over the tropics (Chang, 1968), so farmers in the highlands have an enduring edge over those who farm the semi-arid middle and lower altitudes, quite independently of differences in rainfall amount and variability.[1]

Rainfall Distribution and Variability. Of all the climatic aspects that trouble farmers in the semi-arid zone, rainfall's seasonal distribution and its variability in amount are the two most serious. One way to look at distribution and variability is by rainfall dispersion diagrams (Figures 4, 5 and 6). To take an extreme case, at Kibwezi - Dwa Plantations Ltd., on average the wettest month (November) has 160 mm, but in November 1961 it rained 626 mm (I was there!) the previous year, November's rain was 13 mm. (Plate 6) In months which generally have substantial rain there is tremendous variation year to year.

Figures 4, 5 and 6 serve to show two other facts: that Machakos with its greater rainfall is somewhat more reliable than Kibwezi; and that Machakos and Kibwezi have a bimodal rainfall regime, typical of areas near the equator, whereas Dodoma has a single rainfall season.

A second point about the great rainfall variability of the semi-arid areas is that it is almost totally unpredictable for the local farmer. This month's rainfall does not predict next month's rainfall.[2] In a statistical sense they can be taken as almost

---

[1]Cotton grown in Israel or Egypt will always outyield cotton grown in Uganda or Tanzania by about 40 percent, given optimum practices at both latitudes. Similarly, rice, maize and wheat yields will always be better in mid-latitudes.

[2]Curiously enough this month's rainfall elsewhere does a better job of predicting next month's rainfall at a place. The rainy season in Tanzania develops first in the south and progresses northward. Thus, heavier than normal rains in Mbeya in November may presage heavier rains in later months for areas farther north (Cocheme and Zazzara, 1969).

FIGURE 4

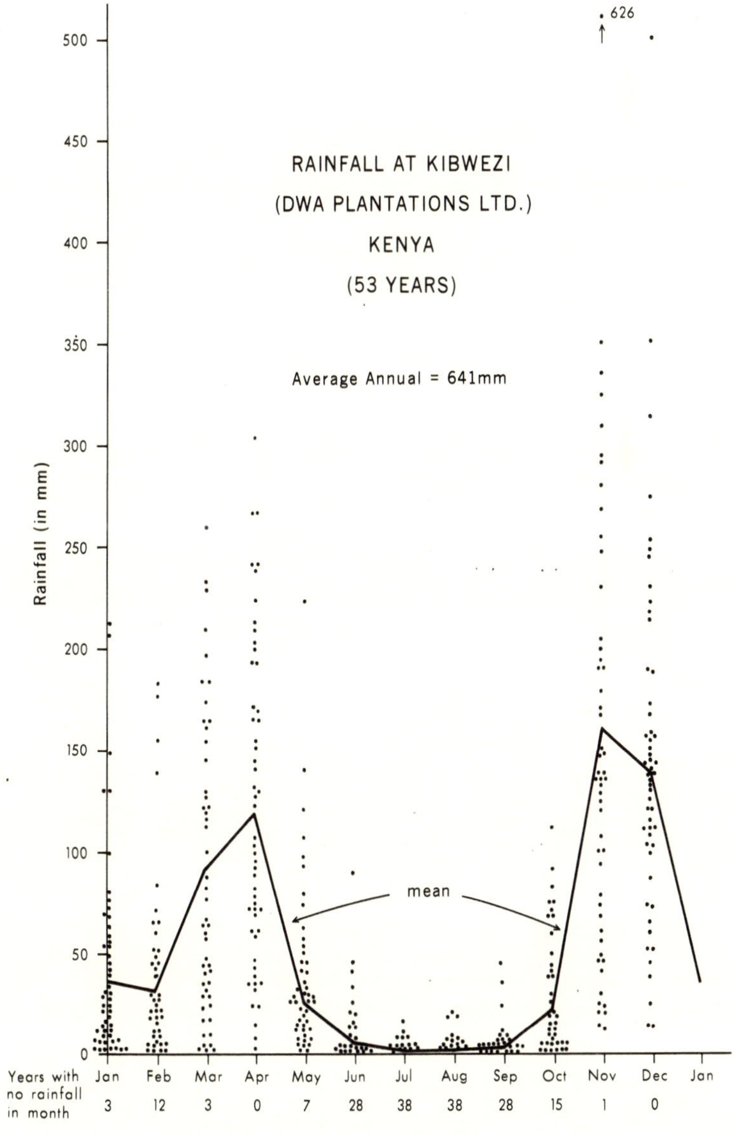
RAINFALL AT KIBWEZI
(DWA PLANTATIONS LTD.)
KENYA
(53 YEARS)
Average Annual = 641mm
Rainfall (in mm)
500
450
400
350
300
250
200
150
100
50
0
626
mean
Years with no rainfall in month
Jan 3
Feb 12
Mar 3
Apr 0
May 7
Jun 28
Jul 38
Aug 38
Sep 28
Oct 15
Nov 1
Dec 0
Jan

FIGURE 5

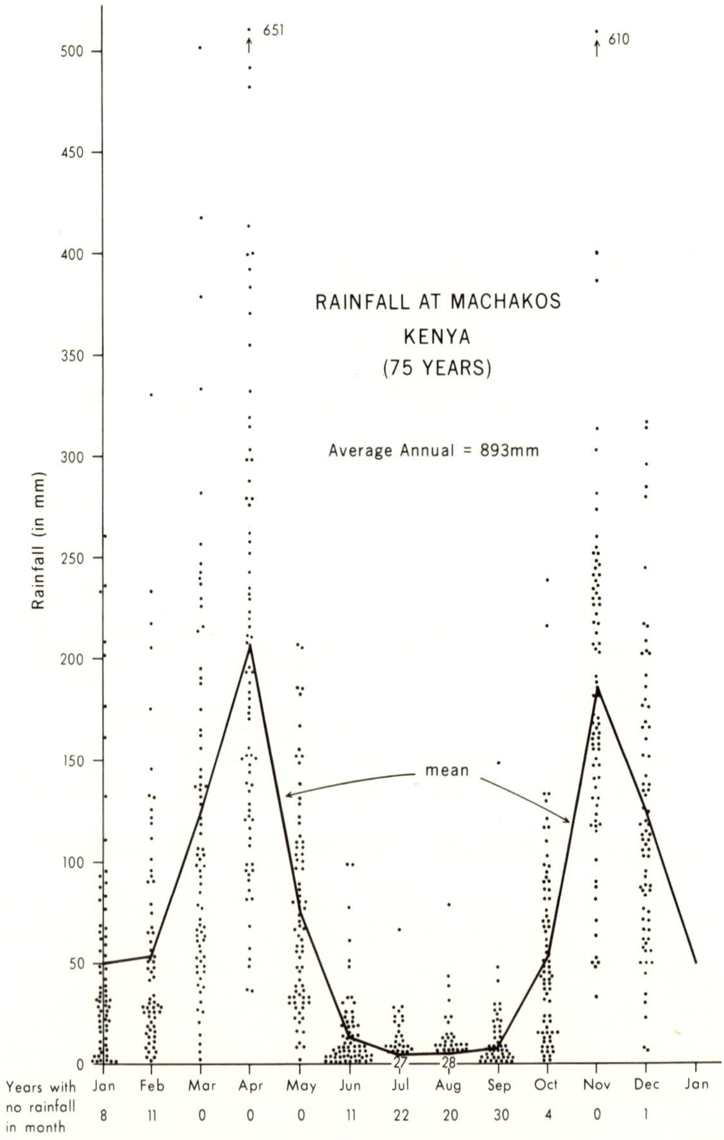
RAINFALL AT MACHAKOS
KENYA
(75 YEARS)
Average Annual = 893mm
Rainfall (in mm)
mean
651
610
500
450
400
350
300
250
200
150
100
50
0
27
28
Years with no rainfall in month
Jan Feb Mar Apr May Jun Jul Aug Sep Oct Nov Dec Jan
8 11 0 0 0 11 22 20 30 4 0 1

FIGURE 6

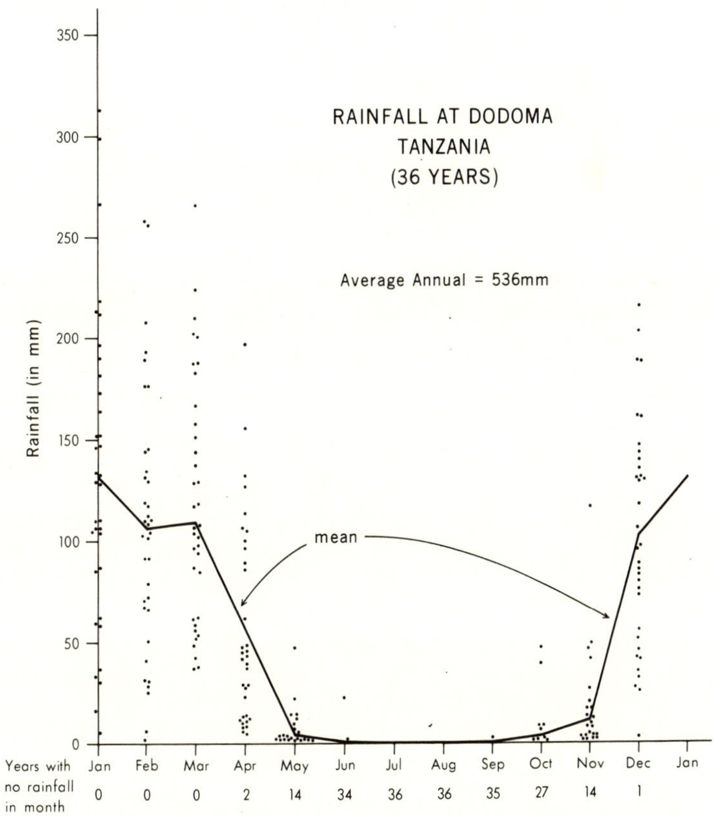
RAINFALL AT DODOMA
TANZANIA
(36 YEARS)
Average Annual = 536mm
mean
Rainfall (in mm)
0
50
100
150
200
250
300
350
Years with no rainfall in month
Jan Feb Mar Apr May Jun Jul Aug Sep Oct Nov Dec Jan
0 0 0 2 14 34 36 36 35 27 14 1

totally independent events. Figure 7 shows the "correlation" by a scatter diagram of the rainfall of the two most significant months of the "grass rains" at Kibwezi--November and December. If we use November's rain each year to predict the rain which will come in December, we get this least-squares equation: December rainfall (mm) = 152.37 - 0.0776 · November rainfall (mm); and we have explained one percent of the variance. Using the equation we can assure the worried farmer that if November's rain is, say an average 160 mm, December's rainfall will be 140 mm, plus or minus 167 mm. That is, we are 95 percent confident that it will be at least zero and not more than 307 mm![1]

There are four general points to make about semi-arid environments: (1) the crucial variable in farming the semi-arid zone is rainfall; (2) rainfall is variable and cannot be predicted in the short run, although long term probabilities can be stated; (3) the choice and management of crops must take into account crucial timing aspects if the crop cycle is to match moisture availability; and (4) the highlands stand with relation to semi-arid areas in an advantaged position in terms of greater rainfall, more reliable rainfall, longer effective growing seasons, and greater potential photosynthesis. We turn now to the East African farmer in the semi-arid zone who transforms this ambiguous climatic resource into a livelihood.

---

[1]Mörth (1967) discovered a fascinating ten-year cycle in rains in the Lake Victoria basin. He has observed that droughts in East Africa are often sudden in onset and that they are system wide, that is, large scale phenomena. Although there may be similar cloud cover from year to year, the amount of rain produced by the clouds may vary greatly.

FIGURE 7

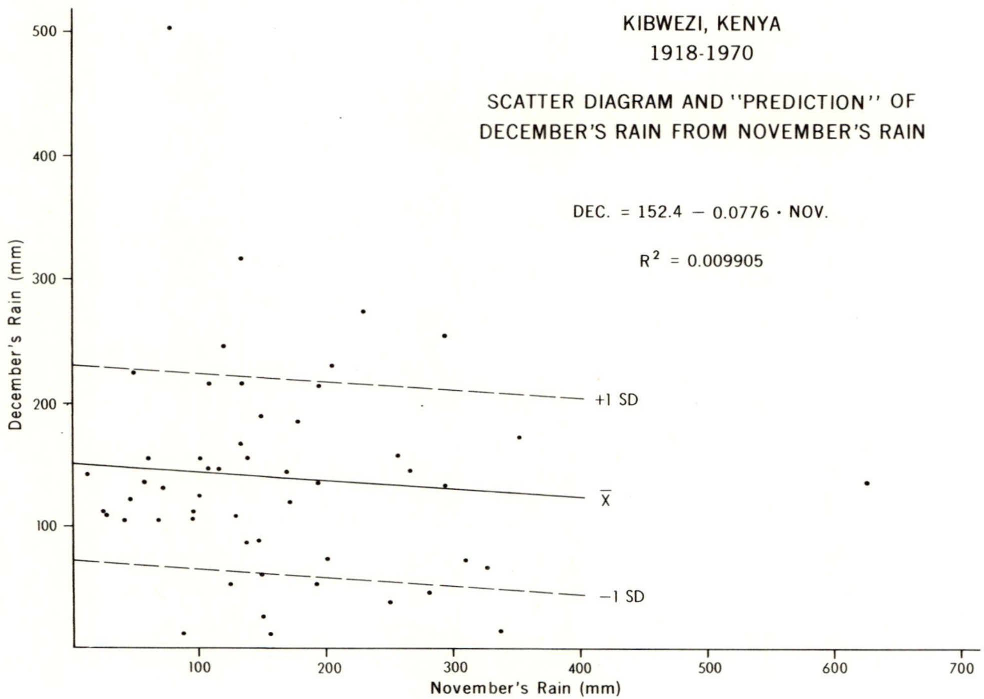
KIBWEZI, KENYA
1918-1970
SCATTER DIAGRAM AND "PREDICTION" OF
DECEMBER'S RAIN FROM NOVEMBER'S RAIN
DEC. = 152.4 − 0.0776 · NOV.
R^2 = 0.009905
+1 SD
$\overline{X}$
−1 SD
December's Rain (mm)
November's Rain (mm)
500
400
300
200
100
100
200
300
400
500
600
700

CHAPTER II

THE WORLD OF THE EAST AFRICAN FARMER

The fact that East African farmers have lived in the semi-arid zone for generations suggests that they know what they are doing and have been successful in making the semi-arid area provide for their needs. Although there is a risk in discussing livelihoods as stereotypes, it will aid our presentation to present a family or lineage agriculture as being typical of the semi-arid areas in the late 19th century, and then to discuss how this mode changed in the colonial and post-colonial periods, particularly through interaction with capitalist agriculture in a variety of settings--global, national and local.

First let us characterize lineage agriculture, paying particular attention to the manner in which it is made to be reliable as well as productive. Those who participate are interested primarily in using what they produce rather than in exchanging or selling it to buy things they need. Of course, exchange of surpluses may take place and even be commonplace, but the farm family produces most of what it consumes. Polanyi (1971). called this _use value_ rather than _exchange value_. Under such circumstances it becomes difficult, if not impossible, to apply conventional categories of economic theory (Chayanov, 1966). Theory based on the interrelationships of wages (of labor), interest (on capital), rent (on land), and profit (of enterprise) cannot be applied to lineage agriculture. To give farm members the dual economic roles of laborer and capitalist exploiting one's own labor superimposes capitalistic categories--an unhelpful fiction. With the removal of wages from the set of relationships, the notions of rent (the farmer is landlord) and profit (the farmer owns the means of production) become quantitatively indeterminate. Thus the pre-colonial farm family's enterprise did not represent capitalistic production but simple commodity production. It has to be judged in terms of what Chayanov called the labor-consumer balance "between the satisfaction of family needs and the drudgery (or irksomeness) of labor" (Chayanov, 1966: xv). The net product (what remained after the farm had been restored "to the same level of production it possessed at the beginning of the agricultural year, i.e., seed, fodder, repairs, replacement of expired livestock and worn out equipment, etc.") could then be divided between consumption, capital investment to raise the farm's potential for production, and savings (Chayanov, 1966: xv). Although it would not be correct to characterize East African farmers as totally subsistence oriented even in the days before colonial rule, they tried, nonetheless, to produce most of what they consumed. Consequently, it is not improper to look at an African society or a subset of it in those early days as a self-supporting, self-contained entity, relying mainly on the products of its own labor.

The present day situation of family/lineage agriculture in the semi-arid zone is, of course, complicated by the colonial experience of the 20th century. People in the semi-arid areas today earn a livelihood by combining many activities. Part of the fruits of their efforts is consumed directly, since people eat the crops they grow and consume products of the stock they herd. Surplus may be sold or exchanged at markets (eggs, milk, stock, grains, fruit). They grow cash crops such as cotton and castor oilseed. Some people process sisal fibers at home and sell them. Many families maintain a line of bee hives in the bush and sell part of the honey and wax. (Plate 7) Some make and sell craft items such as baskets, mats, wood carvings, and bark rope. The brewing and selling of beer is common. Many people hunt, fish, and collect bush fruits. Some poach wild game for meat and participate in the illegal trade in ivory and rhino horn. Many people earn income by cutting firewood and making charcoal, by delivering water and by carrying sand for use in construction. Still others have small shops or are tailors. Most importantly, people sell their labor, both short term and long term, nearby and far away (Figure 8).

FIGURE 8

THE STRUCTURE OF A SEMI-ARID FOOD SYSTEM

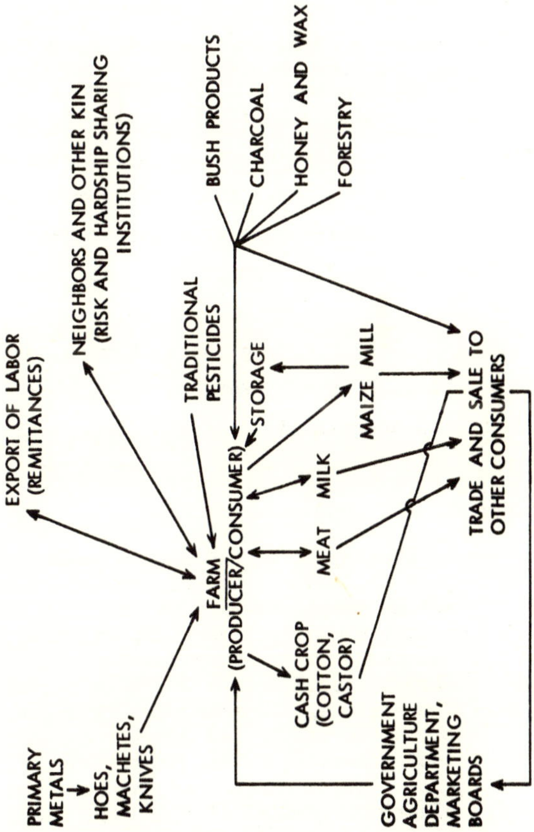

To farm and herd successfully in the semi-arid zone, one must meet certain requirements set by the environment and the nature of crops and animals. These we may call functional requisites (Goldschmidt, 1966). For livestock, the requisites focus on meeting their needs for water, graze, salt and protection from disease and predators. To do this daily, year after year, demands considerable skill and knowledge. One must know a great deal not only about the ability of the animal to withstand physiological stress, but also about environmental management: which grass to save for late grazing, where and when to establish dry season wells so that the stock will be able to withstand the rigors of the daily journey between water and graze. With respect to crops, there are similar requirements of timing and locating activities so that plant moisture and nutrient needs will be met and will result in a good harvest. The system also must take into account the variability of rainfall, it must guard against nature providing too much rain, or not enough, or sending it at the wrong time. Some of the flavor of the rationality and scientific understanding of the lineage farmer's approach to agriculture is captured in a description by Leakey, written in 1934.

> The object of the Kikuyu peasant is primarily to produce food in sufficiency for himself and his family, and he wants not only an abundant food supply but also a varied one.... There are two main wet or 'rainy' seasons when planting takes place, and consequently two harvests each year. Of these two rainy seasons one is comparatively long and one is short, and although some crops can be planted equally successfully during the long and short rains, others cannot.... When a piece of ground has been cleared and hoed and prepared for planting a number of different crops are planted all mixed up on one and the same patch. No attempt is made to segregate the different crops, and no attempt is made to plant in lines.... He plants over the whole area maize, beans of two kinds (beans and Kikuyu beans) and pigeon peas. In planting these the maize and the pigeon-pea seeds are put in first, irregularly all over the plot, and in a few days--when the seedlings have appeared--the two species of beans (i.e. true beans, *Phaseolus vulgaris* and Kikuyu beans, *Lablab niger*) are planted, again quite irregularly, in the

gaps between the maize and pigeon pea seedlings. In a few days the bean and Kikuyu bean seedlings also appear, and then cuttings of sweet potato vine are put in among the growing seedlings of the various other crops.

One of the varieties of bean that is planted (the erect varieties of the true bean *P. vulgaris*) is a very quick-growing plant, and this is harvested long before anything else is ready. This bean is never stored away in granaries for use at a later date, but is consumed more or less at once, and provides the families with fresh food, which--with the sweet potatoes planted at the beginning of the previous short rains, and which are just beginning to yield--is very welcome as a supplement to food made from dried crops from the granaries.

This bean having been uprooted and harvested the plot of ground is now left with maize which is rapidly coming into flower, the tree-peas [pigeon-peas] which are very slowly growing, the sweet-potato which is slowly making a carpet of green vine all over the ground, and finally the slower growing Kikuyu bean. Before long the first rains are over and the hot season starts. The maize is harvested leaving the sweet potatoes, pigeon peas and the Kikuyu bean in possession of the soil. The sweet-potato vine has now completely covered the ground, and it prevents the violent heat of the dry season from sucking up all the moisture from the soil. Even at the driest time of the year if you dig into the soil in a field covered with sweet-potato vine, you will find that it is damp even quite near the surface. Thus aided, the very slow-growing pigeon pea and the slow-growing Kikuyu bean can withstand the hot season and continue to grow. The indeterminate Kikuyu bean goes on flowering and producing a crop for a long time, and can often be seen with dry pods ready to be picked, green pods and flowers all on the same plant at the same time. This crop is also particularly valuable to the Kikuyu as it yields a very succulent form of spinach.

During the dry season very violent thunderstorms, accompanied by an hour or two of torrential rain, are not uncommon. As I have already pointed out the majority of Kikuyu cultivation is on the slopes of hills. The carpet of sweet-potato vine prevents these violent storms from washing the soil down into the stream below, and at the same time it conserves the moisture which results from these storms.

When the dry season is nearly over the pigeon peas are harvested but instead of being up-rooted they are roughly pruned, and left to stand during the succeeding

> short rains when they flower a second time and produce a second crop at the end of the next dry season. The sweet-potato vine also starts yielding a crop just at the end of the first dry season following the rainy season in which it was planted and goes on yielding right through the succeeding short rains.
>
> With the coming of the short or second rains a second plot of ground--which has been prepared for planting during the dry season--is planted very much in the same way as before, only the pigeon pea is replaced by finger and bulrush millet, and meanwhile the plot planted during the previous long rains continues to yield its harvest of sweet potatoes, spinach, and Kikuyu beans (Leakey, 1934).

Intercropping in Kikuyu fields is just one of a large number of social and economic practices which ensure the success of crops, protect the livestock, or share out hardship among a larger kindred. These practices can be divided into three types: use of the site, use of the biophysical situation, and use of the social situation.

The most obvious way farmers in the semi-arid zone protect themselves from environmental risk is to select crops which are adapted to the conditions of the locality.[1] When to plant these crops and how large an area to plant become key elements in the farmer's agricultural plan. For example, suppose Kitemu wa Nguli who lives near Kibwezi (Figure 9) needs approximately 250 mm during either of the two rainy seasons to be successful in growing his crop. Given the choice of farming at place A (which averages 250 mm) or place B, he will choose place B, because the record of seasonal rain there suggests that in the long run he will get the 250 mm he needs 90 percent of the time. Furthermore, since yields in semi-arid areas have a strong direct correlation with rainfall (see Figure 12), Kitemu will plant an area large enough to ensure that if the rains are poor the resulting low yield (2.25 q/ha) will still be sufficient to feed the whole family.[2]

---

[1]This section draws freely from my chapter on "Climate and agriculture in East Africa," 1976, pp. 132-139.

[2]We are assuming a modal household size of five. Kitemu's household is somewhat larger.

FIGURE 9

YIELDS AND AREA PLANTED TO ENSURE A HARVEST EQUAL TO OR GREATER THAN NINE QUINTALS UNDER VARYING AMOUNTS OF RAINFALL (FOR A FAMILY OF FIVE)

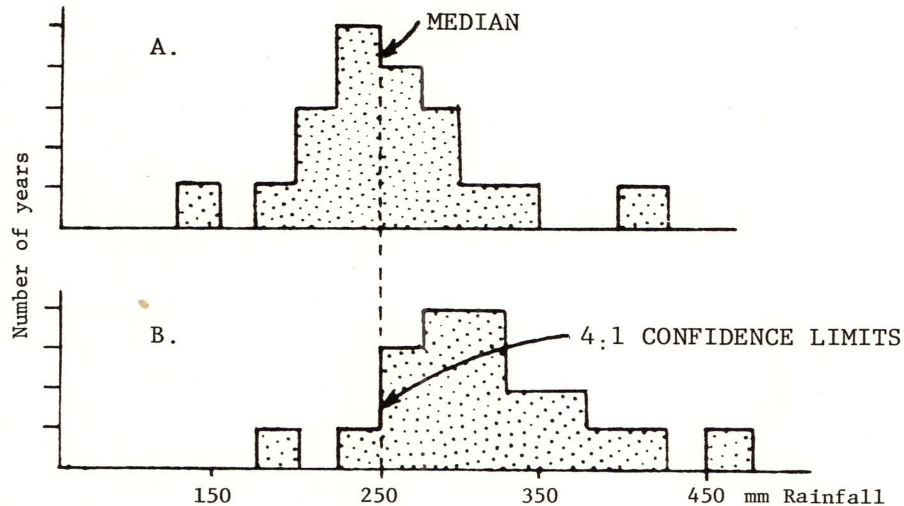

YIELD IN Q/HA	2.25	4.5	9.0
HECTARES NEEDED TO OBTAIN TOTAL YIELD OF 9 QUINTALS	4	2	1

William Allan called Kitemu's practice the "normal surplus" (Allan, 1965). There are, of course, reasons other than the vagaries of rainfall for crop loss. The rains, though totalling more than 250 mm, may be ill-timed. If rains are plentiful, especially late in the season, the crop may suffer from smut, rot, rust or other fungal infestations. There may be outbreaks of locust, attacks by swarms of the Sudan dioch bird (Quelea Quelea Aethiopica). Numerous other agricultural vermin--bushpigs, monkeys, baboons, rats and mice, elephants, buffaloes, hippos, as well as insects of all sorts--are threats to crops both in the field and in the granary (Mascarenhas, 1971).

Nonetheless, if there is a surplus, it can be used in a variety of ways: to sell or trade, to store against a bad year, to brew into beer, to give to others, to celebrate weddings and other occasions with feasting. Of course, in most years a great deal of land must be planted to generate a normal surplus, and as a result land is rested less frequently than it might be if production could be arranged so as not to produce a surplus. In the eyes of a western economist this must seem wasteful of labor, production and soil. In years of really abundant yields, some of the crop may not even be harvested, but will be left in the field to rot or to be eaten by vermin.

To the East African farmer, the system builds in insurance and gives the livelihood greater reliability. Of course, the system only works if a family has sufficient land to plant the appropriate hectarage every year, and enough labor power at two crucial times--planting and weeding--to manage the area needed. These constraints often become crucial for farmers in semi-arid areas.

East African farmers assess the site characteristics of the farm in other ways. They know quite a bit about soil fertility and drainage, and where a particular crop will do best. Farmers can read the vegetation of fallow land and judge its fertility and readiness for use. Most farmers are keenly aware of the importance of correct timing of activities, to ensure optimal use of scarce moisture during the growing season and sunshine during grain ripening periods. Dowker found that each day's delay in planting cost the farmer in Machakos about 5 or 6 percent of potential yield. Thus, a week's delay might reduce the harvest by more than a third (Table 8).

The description of intercropping by Leakey showed that Kikuyu farmers had devised a sound and rational system for cultivation. Their system outyielded an European approach at the time, based on single row-planted crops, by about 25 percent. Further, it protected the soil. Experiences among the neighboring Kamba people permit me to expand Leakey's list of benefits. It should be pointed out that Kamba farmers explained most of these advantages to me, and only a few advantages had to be discovered through agronomic experiments. Since the crops come up all together in a riotous profusion of vines, leaves, and stems, weeds have less chance, and the labor of weeding is reduced. (Plate 8) The phosphorous flush and the nitrogen flush in the soil, which occurs when dry soils are wetted by the first rains, are used to good advantage by all these crops.

> Crops, such as maize, which might be rather indolent about putting down roots and strong tillers in a soil well-supplied with moisture, have to compete with other plants, and thus they put down roots to a depth of several feet. These deeper roots come into play later on in the season. Although the rate of moisture use of all these crops planted together is high, it occurs at the one time of the season when there is likely to be enough moisture available. Further, the interplanting provides a good continuous canopy within which photosynthesis can proceed. Since tropical areas have a relatively uniform radiation income all year around, a continuous leaf canopy, one with a moderate to high leaf-area index, is the most efficient user of radiant energy for photosynthesis (Chang, 1968). Both the farmer's interplanted field and the full-canopied, sometimes multi-storied, natural vegetation of many tropical ecosystems show adaptation to the uniformities of radiation in the tropics (Geertz, 1963).
>
> After about seven or eight weeks some of the crops are harvested. Beans may be harvested as green vegetables. As the season progresses the millet is harvested; beans harvested for seed are taken and the vines pulled up. The number of plants per square meter of soil begins to decline and only crops requiring a longer time to mature are left. The

> bare weedless soil between the plants is dry, which forms a barrier to the movement of moisture to the surface. Evapotranspiration thus is reduced. The sparse plant population remaining is able to tap moisture from a larger volume of soil without competition from adjacent plants. This moisture, combined with the lesser amounts of rain which come at the end of the "grass rains," many times is sufficient to bring the maize and the longer growing millets to harvest. The plants also provide some shade for crops set out to get a start on the main rains. The thick mat of plant cover in the first weeks of the "grass rains" and the second rains also serve to hold the soil. There is also the adaptive fact that if the rains do give out, some crops will have been harvested and the agricultural effort will not have been a total loss (Porter, 1976: 134-135).

The intermixture of crops keeps insect populations and plant diseases at low levels and reduces susceptibility of crops to major loss. The extended harvest period helps to distribute the work load over a longer period.

TABLE 8

YIELD OF TABORAN MAIZE IN MACHAKOS

AS RELATED TO TIME OF PLANTING

Year	Seasonal Rainfall (mm)	Time of Planting	Yield kg/hectare	Reduction in Yield/Day (percent)
Nov. 1959-60	210	Dry	1,979	--
		4 days after rain	1,558	5.3
		7 days after rain	1,170	5.9
Nov. 1960-61	291	Dry	1,526	--
		6 days after rain	954	6.3
April 1962	311	Dry	4,580	--
		6 days after rain	3,284	4.7

Source: Dowker, 1971.

Farmers exploit their situation in several ways, the most common being to disperse fields several miles from one another. Rain storms in the semi-arid areas are commonly small and intense, and a field's prospects for a good harvest may be ensured by a single good rain (Jackson, 1969). Field dispersal is a form of insurance--if the rains fail in one area, perhaps they will be adequate in another. In areas with great local relief, very different ecological zones may be very close to one another (within walking distance) and the people may arrange their residence, land use, and political territory to enable each family to use several differing environments. A good example of this is provided by the Pokot who live in a semi-arid/sub-humid area in northwestern Kenya. The Pokot political territory, called the korok, runs from lowlands (1,000 m) to a high forest zone (3,350 m) and is usually a ridge or hillspur bounded by streams on either side. Several environments suitable for different crops (maize, sorghum, beans, finger millet, white potatoes) and different practices (irrigation), and with differing growing seasons enable Pokot farmers to grow a variety of crops and to spread work of planting, weeding and harvesting out over a nine-month period. A further reciprocal advantage is provided by an economic and geographical division of labor, for the Pokot who live in adjacent semi-arid plains trade livestock and stock products with the hill people in exchange for grain, tobacco and other farm produce. The pastoral Pokot men also say that one consideration in taking a wife, many of whom come from the villages in the hills, is where she comes from. Is it a place with a plentiful and reliable food supply? "It is better, all other things being equal, to marry a second wife from an area whose harvest comes at a different time from that of the first wife's village" (Porter, 1976: 138).

Finally, East African farmers take full advantage of the social "situation," fashioning institutions which insure the family and its activities against environmental risk and creating a system with resiliency and ability to recover from drought and food shortages. The lineage mode of production and way of life incorporates many features which ensure the well-being of all.

Mutual aid is guaranteed by right of kinship, or membership in the same age grade or the same clan. Many rituals and ceremonies are occasions for the slaughter of stock and the sharing of meat and other food among a wide range of kin (Maro, 1975; Taylor, 1974).

The way in which livestock are managed provides one of the best examples of the sharing out of risk. Cattle, sheep, and goats can be moved, traded, and sold. Their numbers increase rapidly in good years, but stock are a comparatively volatile form of wealth and basis for livelihood since they are susceptible to disease, drought, theft, animal predation, and even lightning strikes and other accidents. It is common, therefore, for farmers and herders to split up their herds and have some of them managed in the herds of kinsmen and friends through an often complicated series of livestock exchanges (Goldschmidt, 1976). The terms of the exchange, which has the force of a contract, can vary even within a community, but specify such things as who is to have the cow's milk, blood, manure, and progeny. A time period may be set on the arrangement but more commonly it is open-ended: e.g., until such time as the cow has produced a heifer, who in turn has produced a calf. In addition to the obvious environmental advantages of herd dispersal there are social advantages. It enshrouds in mystery a person's wealth in livestock. In the event of stock theft there will be many injured parties to join in pursuit. Since many livestock exchanges last a minimum of six years, and average closer to ten years, they cement friendships and bind people together, giving them common values and goals. As the Sebei of Uganda say: "Sweet is the cow of debt."

In his study of the human ecology of drought in eastern Kenya, Wisner (1977a) listed 74 actions farmers might take to adjust to drought. Of these, 46 could be regarded as traditional (existing before contact with the industrial agriculture of the west--cloud seeding, sprinkler irrigation, etc.) although some actions, such as seeking wage work on European farms, became traditional only in the last 75 years.[1] The more commonly employed practices are listed in Table 9. This list overlooks, as did the farmers whose responses created it, many details in East African agriculture which are moisture conserving, environmentally beneficial, or risk distributing. By and large the list consists of adjustments rather than adaptations; i.e., they are things a farmer may do when faced with an actual drought, rather than what a farmer can do to the agricultural system itself to make it reliable and resilient.

In this section we have suggested only the merest outline of East African lineage agriculture. As a system it provides at most times an adequate and varied food supply, and in bad times a system for sharing out hardship and loss across the lineage. Those families hardest hit can usually rebuild their livelihood after bad times have ended. The story of how lineage agriculture changed in response to pressures from agriculture operating in the context of capitalism and colonialism is the topic of the next section.

---

[1]An adjustment not mentioned in Wisner's list that used to be practiced more widely was for the family suffering extreme losses to become the client of someone more wealthy. Sometimes a child was pawned or indentured, and thereby became a worker and to a degree a member of the wealthy individual's family. Even in the 19th century there were *abhoi*, or landless people, in Kikuyu country, individuals who worked in return for food and shelter.

TABLE 9

DROUGHT ADJUSTMENTS PRACTICED IN EASTERN KENYA

(N = 595)

Types of Adjustments	Percentage Using the Adjustment
Affect or Know the Rainfall Source	
Pray	93
Pay the rain maker	11
Increase Moisture by:	
Changing Location	
Move cattle	36
Improving Soil Moisture Storage and Distribution	
Cultivate low wet place	48
Dig ridges	33
Irrigate	16
Scheduling for Optimal Soil Moisture	
Plant before first rain	80
Plant dry	71
Plant with first rain	59
Reduce Moisture Need by:	
Eliminating Moisture Waste	
Weed more	64
Quit planting	23
Changing Physiological or Technical Requirements of the Crop	
Plant drought resistant crop	72
Diversity	
Seek wage work nearby	60
Seek wage work far away	40
Hunt or fish	26
Distribute or Share Loss	
Ask help from kinsmen	47
Move to kinsman's farm	47
Ask help from government	21
Send children to kinsman	17
Try to get a loan	15
Ask help from a cooperative	14
Bear the Loss	
Buy food	96
Sell Livestock	53
Collect bush foods	36
Do nothing	1

Source: Wisner, 1977a.

CHAPTER III

IMPACTS ON SEMI-ARID LIVELIHOODS OF COLONIAL AND POST-COLONIAL ERAS

The colonial era for East Africa began in earnest in the last two decades of the 19th century, and although formal colonial control ceased in the early 1960s, its effects are still seen on every hand. This section will sketch briefly, for the East African Colony and Protectorate/Kenya and for German East Africa/ Tanganyika, the effects European settlement and colonial control had on livelihood in the semi-arid zone. Uganda's small areas of semi-arid country present problems for the people who live there, but these problems are not discussed here.[1]

Kenya

Winston Churchill, who passed through the East African Protectorate in 1907 when he was British Under-Secretary of State for the Colonies, described with wonderful prescience the main groups in the young colony which were to contend with one another for control of economic and social policy in the next half century.

> Every white man in Nairobi is a politician; and most of them are leaders of parties. One would scarcely believe it possible that a centre so new should be able to develop so many divergent and conflicting interests, or that a community so small should be able to give to each such vigorous and even vehement expression.
>
> There are already in miniature all the elements of keen political and racial discord, all the materials for hot and acrimonious debate. The white man versus the black; the Indian versus both; the settler as against the planter; the town contrasted with the country; the official class against the unofficial; the coast and the highlands; the railway administration

---

[1]Information on the livelihoods of people in semi-arid areas of Uganda may be obtained by consulting studies by Deshler, 1960; Dyson-Hudson and Dyson-Hudson, 1969; Gulliver, 1955; and Turnbull, 1972.

> and the Protectorate generally; the King's African Rifles and the East African Protectorate police; all these different points of view, naturally arising, honestly adopted, tenaciously held, and not yet reconciled into any harmonious general conception, confront the visitor in perplexing disarray (Huxley, 1935: 1:212).

Until 1963 the views of the European settler group usually prevailed, and they succeeded in establishing the Scheduled Areas, a European highlands comprised of seven million acres (nearly three million hectares) of mostly high-potential land alienated solely for the use of Europeans. One important and unremitting objective was to transform the African population, or at least that portion of it needed, into a wage labor force to work on the farms, estates, and other enterprises owned and run by European settlers. Since lineage agriculture by and large gave Kenyan Africans a self-sufficient livelihood, it became necessary to break into that self-sufficiency and make it necessary for them to supply their labor. This was accomplished by an arsenal of devices, an early one (1902) being the hut tax. Its rationale was explained:

> We consider that taxation is...compelling the native to leave his Reserve for the purpose of seeking work. Only in this way can the cost of living be increased for the native, and...it is on this that the supply of labour and the price of labour depends (Wolff, 1974: 99).

An underlying basis for creating an African labor force lay in land policies. Very early on land was classified into three main categories: alienated land given or sold to European settlers, native land units where the African population lived according to customary rules of tenure, and crown land, whose use or non-use was decided on by government. The Crown Ordinance of 1902 established "jurisdiction over waste and uncultivated land in places where the native Ruler is incompetent whether from ignorance or otherwise, to exercise that jurisdiction" (Quoted in Wolff, 1974: 62). Some of the crown land was subsequently alienated for use by settlers, whereas other parts were incorporated into native land units. Some was held in reserve, and Africans from time to time were allowed to graze livestock there or cut firewood. Land expropriation forced many Africans to leave their homes and to resettle. As part of the Soldier Settler Scheme (1919) in which 2.5 million acres (about one million hectares) were parcelled out as essentially free farms, 64,000 acres (25,000 ha) of the best Nandi land were taken and Nandi were compensated at the rate of six shillings for each house that had to be vacated. In 1921 a rule was passed that all land, regardless of native tenure, remained crown land. Africans became tenants-at-will of the British crown. Thus, Africans found themselves being hemmed in on all sides and unable to use land which formerly had been available.

The settlers used many other means to force Africans to work for them. A poll tax payable by all adults replaced the hut tax in 1903 after it was discovered that people were crowding into fewer houses.

The government regulated and confiscated livestock. For example, the costs of the last punitive expedition against the Nandi (1905) were fully paid for by the sale of captured livestock to settlers. Since the expedition required 12 companies of King's African Rifles and 1,000 Maasai levies, the costs were substantial. The government also undertook forced destocking schemes, justifying such measures as a means of balancing use with the land carrying capacity. These schemes were exceedingly unpopular with the local people. In Ukambani, forced culling of herds was established, partly to supply stock at low prices to a recently opened canning factory (Liebig's Ltd.) at Athi River. A dramatic series of raids, protests, and arrests culminated in a mass march by about 2,000 Kamba men, women and children on Nairobi to confront the governor (Munro, 1975). (Plate 9)

African wages were controlled and the settler population constantly pressed the government to ensure that competition was not permitted. The settlers urged that Kenya's Africans not be allowed to work in the mines in South Africa, that people from Nyasaland, who were used to higher wages, not be permitted to work in Kenya, that Asians be kept out. Africans paid *ad valorem* duties on many imports, while duties on all agricultural implements, livestock for breeding purposes, and commercial seeds and seedlings were exempted. It can be shown that taxes and duties paid by Africans financed the agricultural development of the European farming area (Wolff, 1974: 118). In 1920 a pass system, modelled on one developed a century earlier in South Africa and supported by Kenya's numerous Boer settlers, was introduced.

> Under that system you get a disciplined native; you know where every native is, what his wages are and his employment (G. W. B. Huntingford, cited in Wolff, 1974: 105).

The settlers encouraged Africans to live on their farms and estates as squatters, rather than to work as contract laborers. In return, the European farmer had the right to 180 days of labor per annum from each adult male. In addition, women and children could be called upon in times of peak labor need. Concomitant with all the official and unofficial attention given to the developmental needs of the European highlands was the virtually absolute neglect of the reserves in agriculture and for social services.[1] Africans were prohibited from growing coffee, tea, sisal and pyrethrum, and from selling livestock and dairy products on the European export market. They were paid a lower price for maize than were European farmers.[2] Africans could not

---

[1]One of the most serious manifestations of the near total neglect of African agriculture is seen in the paucity of western scientific understanding of the many crops--staple and non-staple--which traditionally have contributed to the diet of East Africans. This is reflected in Acland's excellent book, *East African Crops* (1971), which devotes 81 pages and includes 223 references to coffee, but gives only 38 pages and 51 references to bananas, cassava, bulrush millet, finger millet, rice, sorghum, and beans--a ratio of 15 to 1 in text and 30 to 1 in references. The 1966 *Review of Applied Mycology* contained 234 references to tobacco plant diseases but only two to cassava diseases.

[2]European farmers were subsidized to grow maize. The Food Shortage Commission of Inquiry reported in 1943: "Since the Colony cannot safely depend on native grown maize to satisfy its

set up a shop without a government license. They had no access to credit. On the other hand, the European farmers were aided at every turn. They acquired land cheaply, with 999 year leases. Agricultural research and extension was geared wholly to their needs. They were given tax breaks on imported equipment and exported crops, and the rail and road infrastructure served the European highlands. Finally, the government aided farmers in labor recruitment and in provision of credit.

Meanwhile, the African population grew. Population densities in rural parts of the reserves reached extremely high levels, 500 and 600 people/sq. km being commonplace in parts of Kikuyu country and in Machakos District, less than a fifth of a hectare per person. There was tremendous land fragmentation and landlessness, although in Kikuyuland there had been a landless group long before the arrival of European settlers. At its peak in 1960, the European highlands had 3,600 Europeans in agriculture, overseeing an African labor force of 268,000, about 75 workers per farm (Table 10). The average value of commercial agriculture to each European in agriculture in 1960 was $29,000. African incomes from commercial agriculture (that is, the same economic domain with European farmers) were $3.50/capita.[1] The rate of profit in 1960, taking into account replacement of capital investment, loan repayments, and the wage bill was over 130 percent.[2]

---

internal requirements, it must look to the European farmer to produce maize. The European farmer cannot afford to produce maize for sale unless he obtains a return at least equal to that obtainable from other farming activities" (_Food Shortage Commission of Inquiry Report_, 1943: 46, cited by H. Karani, 1965). The Increase of Crops Ordinance, 1942, whose provisions were continued in the 1955 Agriculture Ordinance, granted European farmers a higher price for maize than that which African farmers could hope to get (Karani, 1965).

[1]The value of what Africans produced and even traded on local markets should not be confused with their participation in the commercial circuits in which European farmers traded--the national and international markets. In national and international trade, the value of African produce in 1960 was about $3.50/capita.

[2]An analysis of the "prices of production" (see Emmanuel, 1972, Chapter 2) was made of European agriculture and industry in

The end result of this differentiation between white highlands and reserves, between European and African, was Mau Mau, the Swynnerton Plan, and in 1963, independence for Kenya. The Swynnerton Plan to intensify African agriculture was intended to make the African reserves produce wealth. This was to be accomplished by consolidation of land into economically viable holdings, registration of freehold titles, and removal of the ban on growing coffee and other cash crops. (Plate 10) Swynnerton said: "Former government policy will be reversed and able, energetic or rich Africans will be able to acquire more land and bad or poor farmers less, creating a landed and a landless class. This is a normal step in the evolution of a country" (Swynnerton, 1955: 10). Whether or not the creation of landed and landless classes is a "normal step" can be challenged, but undeniably it is occurring in Kenya. As of 1977 there were more than 300,000 landless families in Kenya, or about 1.7 million people.

It is difficult to separate pre-independence social and economic processes pursued by the colonial administration and the European settlers from those which have been followed since Kenya gained its political independence in 1963. The policies of the earlier era created a momentum which carried on into the post-independence period despite the fact that different people were making decisions. The colonial policies set in place a modern capitalist form of agriculture supported by an extensive infrastructure of cooperatives, marketing boards, credit-granting institutions, research and extension services, and roads. In the post-independence period European settler farms in the Scheduled

---

Kenya as of 1960, using economic survey data and agricultural census data for the Scheduled Areas. Calculations for the total wage bill, capital investment, loan repayments, taxes, other non-labor costs of production, gross value of sales, and net value of sales were made. The rate of profit in 1960 averaged 138 for agriculture and 10 for industry. The profit in agriculture may have been unusually high in 1960 in comparison with rates in the 1940s and 1950s. Overall labor costs (that is, average individual wage differentials) for Europeans, Asians, and Africans employed in European agriculture were in the ratio 30:13:1.

TABLE 10

SELECTED FACTS ON KENYA'S AGRICULTURAL ECONOMY - 1960

Total European Population		61,000
Number of Europeans Engaged in Agriculture		3,600
Income From Sale of Agricultural Produce (in K£000)		
African		9,624
Non-African		37,870
Employment on Non-African Farms		
European		1,818
Asian		800
African		
Male	171,800	
Female	67,300	
Juvenile	29,200	
Total		268,300
Per Capita Income from Commercial Agriculture		
African		$ 3.50
European		$ 29,454.00

Source: East African Statistical Department, 1961.

Areas were transferred to African ownership in a variety of ways. Over three-quarters of the large-scale farms were transferred intact to Kenyans with sufficient means to purchase them (Odingo, 1971). Many people in government service took over these farms. By the late 1960s some 35,000 families had been settled as smallholders in the Scheduled Areas. In many instances they had previously been landless. The people of the semi-arid areas were marginal in environmental, economic, and political terms prior to independence. Since 1963 their status has not improved significantly despite governmental efforts to increase productivity and improve stock raising. The 1974-1978 Development Plan gave considerable attention to problems people have in semi-arid areas, through the work of the Marginal Lands Committee of the Division of Land and Farm Management, Ministry of Agriculture, and through smallholder and integrated rural development projects. These projects are jointly undertaken by the Government of Kenya and various aid donors--United States Agency for International Development, United Kingdom Ministry of Overseas Development, Food and Agricultural Organization/United Nations Development Programme, World Bank, Arab Bank for Economic Development in Africa, and the European Economic Community. The Development Plan for 1978-1983 probably will place even more emphasis on marginal and semi-arid lands (Palmer, et al., 1978: 1).

In the period 1920 to 1955 the combined effects of colonial policies and rapid population growth increased pressure on resources in the native land units generally. The high potential areas were capable of agricultural intensification and increased productivity; these became the main target of development under the Swynnerton Plan which commenced in 1955 (Davis, 1977). The plan took an integrated approach to rural development. Besides concentrating on African production of cash crops, it addressed underlying questions and effected land consolidation and awarded permanent tenure to individual farmers. It also provided for "agricultural extension staff, farmer training, rural water supply, cash crop marketing, credit facilities, and the upgrading of livestock" (Davis, 1977: 38). Although no longer called the

Swynnerton Plan, the program pursued by the government since independence in the former native land units and in the former Scheduled Areas has had the same objectives and has been based on many of the same premises. Where it differs is that government has abandoned the close supervision (and consequent modest pace) which characterized integrated rural development in the late 1950s. Production goals have been placed ahead of welfare goals, and are reflected in the increase in the proportion of export crops grown by smallholders. In 1957 smallholders produced under 12 percent of four main exported crops (coffee, cashews, pyrethum, and tea), but by 1971 the proportion had increased to nearly 50 percent.

The post-independence years have witnessed two processes--the peasantization of the Scheduled Areas on one hand, leaving many of the larger farms intact and in the hands of individuals, and the extension of the Swynnerton European Highlands model (credit, farm size, use of technical inputs) to the high potential former reserves. The whole highlands, not just the former Scheduled Areas, have become a developing core following a market-oriented private entrepreneurial path of development. The semi-arid areas have continued to play the role of periphery--the location of landless migrants who try to farm areas drier than they are experienced with, the source areas of male migrants who work for farmers in the highlands.

In addition to the environmentally disadvantaged position of people in semi-arid areas, there are several economic and political disadvantages. Land consolidation in highland areas made farmers eligible for loans, which increased their ability to hire labor. With profits from coffee, they began to accumulate middle potential and irrigable land in lower areas, by buying out farmers in the sub-humid and semi-arid areas. In Meru-Tharaka, land purchase and the creation of the new Meru Game Reserve took away both wet season and dry season grazing land from the local people, but they were powerless to do anything about it because political power is also in the hands of the wealthier farmers.

People in the semi-arid zone have not only lost control over land, they have also lost control of their own labor in the sense that the daily wage, the wholesale price of commodities the farmers of the semi-arid zone sell, and access to credit for development loans are all influenced or controlled by wealthy farmers in the highlands.

To quote Wisner's study of drought in Kenya:

> The entire structure of the agricultural economy including such sub-systems as finance, research, extension, marketing, and patterns of drought adjustment [is] tailored to maximize the profits and range of choice of the biggest farmers. Since farmers big and small 'compete' for the same capital in the form of credit, extension services, veterinary aid, etc., the big farmers' gain is often the small farmers' loss.... Public credit to a big farmer to buy overhead sprinkler irrigation draws down the fund of resources that might have gone to small farmers. In 1971-72, K£3.1 million in loans under the Guaranteed Minimum Return Seasonal Crop Credit Scheme (GMR) [a crop insurance scheme] went to 5,500 farmers only.... [To be eligible farmers must have more than 15 acres (6 ha) planted to maize or wheat.[1] In the Kikuyu high potential areas at the time land registration was completed, about 1970, 86 percent of all holdings were smaller than 7-1/2 acres (3 ha).] The same public corporations which buy the large-scale farmer's maize also buy the small surplus of the poor peasant. Institutions that supply credit to the one group influence the definition of 'credit worthiness' of the other (Wisner, 1976: 38, 46; see also Leys, 1974).

Title to land has been one criterion of credit worthiness. Registration of land is virtually complete in many parts of the highlands, but hardly started in the semi-arid zone. Credit advanced under the Guaranteed Minimum Return program provides up to K.Sh. 180/acre (K.Sh. 444/ha) for seeds, fertilizer, pesticides, contract cultivation, fuel, machinery repairs and spares, bags, harvesting and transport, and insures the farmer if the crop fails from natural causes.

---

[1]Government policy may be changing. The Co-operative Bank planned in 1976-77 to channel some K.Sh. 1.2 million to smallholders, and World Bank support of an Integrated Agricultural Development Project was also designed to provide smallholders with credit, equipment, seeds, fertilizers and extension services (Legum, 1977: B243). How smallholder was defined in these programs is not clear.

Any surplus the lowland creates is marketed through the highlands, and is often handled by middlemen from the highlands who buy cheaply and add a big markup to the price when they sell. At times of good harvests in the lowlands, better yields in the highlands ensure that low prices prevail, thus the farmers get little when they have something to sell. In times of food shortages, farmers in the semi-arid areas must buy from the highlands at high prices. To pay for this they may sell part of their land, but more likely they will sell their labor, travelling to the highlands to seek work. The large number of hungry migrants ensures that their labor can be purchased for much less than it is worth. Kenya has long maintained the price of a day's labor (about 6 hours--all morning up to 1 or 2 PM) at K.Sh. 2.[1] When food shortages cause many to seek work, the lowland migrants are paid K.Sh. 1/day or less, which is not even a subsistence wage. In addition, migrancy, especially long term, seriously disrupts family life, and jeopardizes the well-being of children.

A continuing problem of semi-arid areas is that there are usually better areas nearby, where investment of scarce capital will create a greater economic outcome. Faced with the choice of where to invest, decision-makers will usually opt for the greater economic outcome. I have dwelt on the history of Kenya and the relations of highlands with the semi-arid areas to show that problems people in semi-arid areas have are not merely _in situ_ and local either in cause or consequence. Those who live in the semi-arid areas of Kenya are not a people apart; they are a marginalized part of a larger system.

Tanzania

Tanzania is especially interesting in that it permits comparison with Kenya, both during their colonial days and since

[1]Inflation in recent years has increased these figures. In 1976 the established legal minimum wage was K. Sh. 7.00/day, but surveys showed that in fact "the rates actually paid...averaged only Sh. 4.2 (Eastern Province); Sh. 3.0 (Western); and Sh. 3.5 (South Nyanza)" (Rusch and Rose, 1977: 7, 36).

independence. We can divide the history of Tanzania into three periods: before 1884, the period of German and British colonial rule--1884-1961, and the period since independence. Prior to the arrival of the German colonialists, lineage agriculture was almost universal, and self-sufficiency the rule in most things, aside from salt and iron (Kjekshus, 1977a). In portions of Tanga Region (Muheza, Usambara, Uzigua), the people exported large quantities of food to the slave-worked clove plantations on the islands of Zanzibar and Pemba. Early German travellers, for example, Baumann, often commented on the general prosperity and well-being of the people in the areas they traversed. In 1890 Baumann described Mgera, which lies within the semi-arid zone in western Tanga Region, as follows:

> The Wazigua produce considerably more grain than they need for their own consumption. As a result they export grain in very considerable quantities to Sadani and Pangani. Because of this continuous trade with the coast, which is carried on in addition by numerous Swahili traders who pass through the land, the door has been opened wide to the import of European goods and the spread of the culture of the coast.... In Mgera everyone wears fine, white cloth, muskets and powder are common and almost every child speaks Swahili (Iliffe, 1971: 10).

It would be incorrect to conclude that life in the semi-arid zone or anywhere in East Africa, for that matter, was without problems. The rains failed then as now, sometimes there were swarms of locusts or army worms, and food shortages, stock losses and occasional famines occurred. Indeed the decade of the 1890s at the time of early colonial penetration stands as one of the worst on record. It reads like a litany on bad times--the small-pox epidemics of 1892 and 1898, droughts of 1891, 1892, and 1896, locusts in 1894, 1895, and 1899, rinderpest epidemics in 1890, 1892, and 1894, and influenza in the period 1899-1901. The havoc created by these events probably made it easier for European settlement to take hold in the Kenya Highlands.

Table 11 lists famines and food shortages as remembered by the Gogo people. Although the memory of droughts and food shortages may be faulty for the earlier part of the record, examination of the table seems to show an increased vulnerability and an increased frequency of bad years since the 1930s. In the absence of climatic change, the increased vulnerability of the Gogo must come from other sources.

During the colonial period in Tanzania three sorts of social and economic differentiation or inequality developed: (1) between urban and rural areas, wages being higher in urban areas; (2) between rural areas, since some developed rapidly and others did not; and (3) between individuals, largely through differences in participation in cash crop production. In the semi-arid parts of Tanzania cash crops did not develop except for peasant-grown cotton in the Mwanza/Shinyanga area. Livestock were the most important cash crop, and they had had that role long before the arrival of the Germans.

A further change, which affected the semi-arid areas in particular, was the acceptance of new food crops. Bulrush millet and sorghum, the traditional staples, are quick growing, drought-resistant, nutritious, and store and transport well. The past 150 years have seen the diffusion and adoption of rice, maize and cassava as staples. In the semi-arid zone, maize in particular has become a staple crop, displacing millets and sorghum. Although maize has the advantage of better yield when rains are good, and of being less susceptible to bird damage, it is less drought-resistant than sorghum or bulrush millet, more vulnerable to insect damage, and does not store as well.

European settlement never achieved a dominant position in Tanzania as it did in Kenya. At most one percent of the land was alienated, and that was mainly in sisal plantations in Tanga and Morogoro regions along rail lines. Thus, many Africans at an early date were able to grow cash crops. But crops like coffee, tea, and cotton cannot be grown everywhere, and thus a three-fold differentiation emerged: (1) European plantations and estates which employed Africans, the bulk of whom travelled considerable

TABLE 11

FOOD SHORTAGES - UGOGO, TANZANIA

1870*	D	1897		1924		1951+	D
1871		1898		1925*	D	1952+	SD
1872		1899		1926*	D	1953*	D,AW
1873		1900		1927		1954*	D,AW,L
1874		1901		1928		1955*	F,L,HR
1875		1902		1929*	LR, L	1956+	SD
1876		1903		1930*	LR, HR, F	1957	
1877		1904		1931+	HR	1958+	SD,HR
1878*	D	1905*	L, R	1932+	L, LR	1959	
1879		1906		1933*	LR	1960*	D
1880		1907		1934*	D	1961*	D,HR
1881*	SD	1908		1935*	D	1962*	HR
1882		1909		1936+	HR, F	1963+	I
1883		1910+	D	1937+	D, R	1964	
1884		1911		1938*	LR	1965+	D
1885*	D	1912		1939+	LR	1966*	D
1886		1913		1940		1967+	D
1887		1914		1941*	D	1968	
1888*	L	1915		1942*	D, SD	1969*	D
1889		1916+	T	1943*	D	1970	
1890*	L, R	1917*	T	1944		1971	
1891*	L, R	1918*	T, D	1945*	LR	1972	
1892*	L, R	1919*	Inf, D	1946*	D	1973*	D
1893		1920*	Inf, D	1947*	D	1974*	D
1894		1921+		1948*	D	1975*	D
1895		1922		1949*	D, R		
1896		1923		1950+	D		

Key: Magnitude of shortages:
+ = serious
* = extreme

Causes of shortages:

AW = army worms
D = drought
F = floods
HR = heavy rains
I = insects
Inf = influenza
L = locusts
LR = late rains
SD = Sudan dioch
R = rinderpest
T = troops (WWI)

Source: Brooke, 1967a, 1967b; Rigby, 1969; Patton, 1971; Ndorobo, 1973; Mascarenhas, 1975. There is some disagreement among these sources.

distances to find employment, (2) African areas where farmers grew cash crops--coffee, cotton, tobacco--for the world market, or food crops for local sale (for example, the Bondei of the Tanga lowlands provided food for the laborers on the sisal estates, many of whom came from western Tanzania and Burundi); and (3) areas of labor outmigration and rural stagnation. The semi-arid areas fall generally within the zone of stagnation and outmigration, with cattle as the main export.

Differentiation did not reach the extremes evidenced in Kenya, but at the time of independence there were clearly "islands" of development where cash crops were grown and groups of individual farmers had become wealthy. In the semi-arid wedge, these islands were Kilimanjaro and Meru (coffee), Mbulu (wheat), and Sukumaland (cotton).

In the 1950s British policy on agricultural development in Tanzania, as in Kenya during the Mau Mau, concentrated on the making of progressive farmers who would have freehold title to land. Independence was on its way, and some colonial officials hoped to leave behind a class of westernized wealthy farmers to "provide stability" in rural areas. One official quoted by Iliffe stated:

> I want to see the emergence from our hitherto undifferentiated African society of a substantial number of rich men.... I would like to see men in a sufficiently strong financial position to be able to send their sons overseas for education, to afford motor-cars, good houses and the like, and I believe that the emergence of such relatively wealthy individuals in the community will provide a stabilising factor of immense importance to the future of this country (Iliffe, 1971: 37).

By independence, some class differentiation had occurred. The policy of Julius Nyerere and TANU (Tanganyika African National Union), even before independence, had been to stop the growth of class divisions and economic inequality. Since 1961, Tanzania has attempted to follow a socialist course. President Nyerere felt that with 90 percent of the population living in rural areas, the best way to improve the well-being of most citizens was through rural development. Government policy has consistently tried to prevent the growth of extremes in wealth among regions and among people. In 1962 land was nationalized and all freehold land became leasehold. A pivotal event was the Arusha Declaration, published February 5, 1967 (Nyerere, 1967a). It described the goals of socialism, gave special guidance to the leadership, and brought under public ownership all financial and monetary enterprises and the most important commercial and industrial concerns. The government nationalized (with compensation) banks, insurance companies, food processing and certain manufacturing firms.[1] The eight largest import-export wholesaling enterprises became the nucleus of the State Trading Corporation. The government took controlling shares of major manufacturing industries and the sisal industry.

In 1968 a policy of Ujamaa Vijijini (roughly, familyhood in the villages) was adopted for the transformation of the rural areas, and it is intended eventually to apply the idea to urban neighborhoods. Ujamaa is a way to mobilize labor and help people through education and practice to live in a socialist way. The ujamaa village is a nucleated settlement for which it is more economic to provide such things as safe water, a primary school, and a dispensary, than for an area of dispersed settlement. Ideally, the land is held communally, labor is contributed, and the proceeds of the harvest or other economic activities are shared out equally. Decisions on investments of proceeds (for a maize mill, a tractor) are taken by adult members of the village through discussion and vote. In practice, each family retains some land to cultivate for itself, usually one acre in the ujamaa village, and often an older farm where the family lived before joining the new village. People also contribute labor to the

---

[1]When the banks were nationalized the three main banking firms (Barclays DCO, Standard Bank, and National and Grindlays) "withdrew all their European personnel within a period of two months, and also destroyed the instruction manuals with descriptions of banking procedures" (Van de Laar, 1972: 84).

ujamaa farm, whose proceeds are used to finance village developments (deSouza and Porter, 1974: 66).

Winans provided a succinct summary of the more important objectives and policies pursued by the government of Tanzania in recent years.

> Tanzania is not attempting to abandon the larger scale of the social order that was created in the colonial period. It is experimenting with ways of counteracting the tendency of such a larger system to extract wealth along the various kinds of clines [in cost and productivity] that exist in any large system. There is a deliberate attempt to block the extraction of labor from the countryside through efforts to halt urban migration. There is a policy of channeling development investment to the rural areas and denying it to the urban places in order to halt the subsidization of the towns by the rural poor. There is a conscious decentralization of decision-making within Government to allow the expression of local demands to be felt. There is an effort to increase the size of rural places to allow them to achieve the benefits of "pooling" that come from this increase in scale of production and consumption. There are deliberate ceilings on salaries and wages to hold down growth of wealth differences, and there are limits placed on farm sizes to keep income and capacity to hire others to do one's labor within rather strict limits (Winans, 1978).

Since 1972 a policy of radical decentralization of administrative functions to the regions has been pursued. Decentralization was carried out to make planning and development officers more responsive to the needs of people, and so that local citizens could more effectively participate in planning and development efforts. It is agreed that "development is a multi-faceted process that requires joint efforts in agriculture, health, education, population, nutrition, water, etc. Emphasis on agricultural development alone has often resulted in benefits going almost exclusively to large and more innovative farmers" (Blue and Weaver, 1977: 9). Thus, regional planning replaced a national sectoral approach which formerly dominated administration and economic planning.

Enough time has passed that evaluations of *ujamaa vijijini* have begun to appear, yet it is still probably too early to assess the permanent effects of this program on the economic and social life of Tanzanians.[1] After all, the big move of millions of people into nucleated settlements was accomplished only in late 1976. Evaluations published thus far tend to be pessimistic and to reveal the ideological predispositions of the authors.

Ujamaa, according to Frances Hill, is an "ideology of production" (Hill, 1975: 217). Its basic premise is that by coming together to live in nucleated villages, Tanzanians can increase their productivity, make improvements in agricultural techniques, and realize economies of scale both in production and in the provision of social services. Increased production will mean increased income and a higher standard of living. In ujamaa, a socialist ideal is supposed to govern the relations of people with one another and with their system of production. In true ujamaa, village work is communal and the product of communal effort is shared equally. The ujamaa settlement makes it easier to exchange information, to discuss matters of common concern and decide things democratically, to help adults learn to read and write, and to increase the political consciousness of its members. Further, practice of ujamaa will prevent the emergence of a class society and strong economic differentiation.

The final item in TANU's creed, which begins the Arusha Declaration, stated Tanzania's socialist goal.

> [It] is the responsibility of the State to intervene actively in the economic life of the nation so as to ensure the well-being of all citizens and so as to prevent the exploitation of one person by another or one group by another, and so as to prevent the accumulation of wealth to an extent which is inconsistent with the existence of a classless society (Nyerere, 1967a: 1).

A transformation of the sort envisaged by the program of ujamaa was bound to face great difficulties. In the first place,

---

[1] Recent assessments of ujamaa and other developments in Tanzania are Blue and Weaver, 1977; Campbell, 1975; Hill, 1975; Huizer, 1978; Kjekshus, 1977b; Lofchie, 1976; McHenry, 1976; Omari, 1976; Pratt, 1975; Shivji, et al, 1973; and Shivji, 1976.

it called for reemphasis of a communalism in traditional culture which in some instances either was not there to begin with or had been severely modified during the colonial period.

> We, in Africa, have no more need of being "converted" to socialism than we have of being "taught" democracy. Both are rooted in our past--in the traditional society which produced us (Nyerere, 1967b: 170).

The pursuit of individual goals was far more developed in many areas, particularly where cultivation of cash crops has been important (Kilimanjaro, the Usambaras, Sukumaland, West Lake, Tukuyu). Paul Maro (1974) documented the decline of sharing and communalism among the Chagga of Kilimanjaro. By the time _ujamaa_ was introduced, differentiation and emergence of class differences had gone quite far in certain areas, for example, among the maize growers of Ismani (Awiti, 1972). According to Shivji (1976), before independence there was a metropolitan bourgeoisie, mainly British, a commercial bourgeoisie, almost wholly Asian, ranging from estate owners and large wholesalers to the rural shopkeeper, an African petty bourgeoisie consisting of teachers, wealthy farmers, civil servants, military personnel, and some traders and shopkeepers, a few yeoman farmers, workers (mainly migrant), and peasants. Differentiation had progressed far enough that these groups would emerge to pursue their self-interest along class lines in the post-independence period. Furthermore, some authors assert that because land is plentiful in Tanzania, it was "possible to bring families into collectives, _Ujamaa_, without dispossessing an entrenched political and economic class" (Campbell, 1975: 47).

The gap between the idea of _ujamaa_ and attempts to implement it is large. Governmental writing about _ujamaa_ speaks of it as coming through initiatives at the grass roots, yet in fact, much of the pressure for the creation of new villages came "from the top--down." There was administrative ineptitude and poor physical planning in selection of sites and village layout, and once established the relations of villagers with government officials were reminiscent of colonial times. Some residents of new villages thought they were now working for the government and referred to the new settlements as Nyerere's villages. Many villages were *ujamaa* in name only and had no socialist or cooperative features.

It appeared that those who were self-reliant and already making a good living were by-passed by government. Aid to *ujamaa* villages avoided the developed, high potential parts of Tanzania in favor of areas of pioneer settlement and the creation of new villages (Rald, 1970). It concentrated on marginal environments such as Dodoma and Kigoma, where the provision of water and social services might be seen as sufficient attraction to persuade poor people living in rural dispersed settlements to come together in one place. In Mtwara, Ruvuma and Kigoma Regions some settlements were created to accomodate refugees from Mozambique and Burundi. The actual provision of expected social and physical services was crucial to the permanence of *ujamaa* villages, and where they ceased or failed to materialize, settlers often returned to their former homes.

In many areas people were reluctant to make the major shift to new settlements and did so only when forced. Many farmers tried to maintain both old and new home until it was clear that the new one would suffice for their needs.

> They had moral, social and economic attachments to their old area of residence. In comparison to [those who moved voluntarily], they were in the majority. [This group was] generally comprised of people with permanent crops and big private gardens, cattle, goats, sheep and better houses (Omari, 1976: 126).

Certain groups in Tanzania, the Sandawe for example, maintain a dispersed settlement pattern for honored social and economic reasons, and some authors have predicted an actual worsening of social, economic, and environmental conditions where *ujamaa* villages have been imposed on an unwilling populace (Newman, 1973; Kjekshus, 1977b).

In economic terms the *ujamaa* experiment has not yet proved to be a success. The purpose was to increase agricultural productivity and total production, but since 1968 the agricultural sector has barely kept pace with growth in population. In the first place actual investment in agriculture between 1966 and 1973 was only 7.1 percent of fixed capital formation (Ministry of Finance and Planning, 1964-1972). During the same period producer prices were increased little because of risks that well-to-do farmers would gain most of the benefit of increased prices, thereby differentiating rural incomes further. Through ujamaa and "villagization," Tanzania's rural, agricultural people bore the brunt of the major institutional changes in the country. They had to move, to abandon fields and farms and devote time to making new lives, which meant building new houses, clearing new areas for fields, and so forth. A considerable amount of wealth and productive capacity was necessarily abandoned in making the move. There were also inevitable delays in making a new area as productive as the old. In addition, problems were brought on by the devastating droughts of 1973 and 1974, the attendant food shortages, the need to spend large sums of foreign exchange for grain purchase and transport, the quadrupling of oil prices, and a worsening in the terms of trade for Tanzania's main export crops. These expenses took nearly all of the money intended for development in these years.

Since 1975 the government has taken several steps calculated to improve rural livelihoods. At the same time its pragmatic approach risks an increase in economic and class differentiation. Although villagization has been carried out, it is the physical settlement pattern which has been transformed. The term ujamaa is no longer officially used and the communal aspects of ujamaa have been suspended for the time being (Legum, 1977). Some services (health, education, water) will be consumed communally, but production is to continue on an individual basis. Government policy will be to pay higher prices to producers of agricultural commodities as a strategy to increase production. Hill (1975) and others fear that this is simply a strategy to induce production of an exportable surplus which the governmental elite will be able to capture for its own purposes.

There has been a further shift toward using a higher proportion of development funds for directly productive purposes rather than for social and economic infrastructure (water, power, communications, education and health). This change in investment strategy is reflected in two figures. In the period 1971/72-1973/74 annual investment in directly productive activities was 210 million shillings, which represented 22.9 percent of government expenditure. The 1976/77 budget provided for 1,317 million shillings for directly productive purposes, or 43.3 percent of government expenditure. Between the two dates, the amount going to agriculture increased five-fold, and the proportion going to agriculture increased by nearly 50 percent.

In early 1977 President Nyerere published a remarkable evaluation of Tanzanian accomplishments and disappointments in the decade since the Arusha Declaration (Nyerere, 1977). In his view Tanzania in the past ten years had stopped the drift toward a class society, established some of the attitudes and institutions necessary to the development of socialism, made reasonably good progress towards providing basic health, education and transportation facilities for all, and contributed to the freedom struggle elsewhere in Africa, thus helping to ensure its own freedom. Tanzania has avoided the creation of a landless class, the evils of private usury, and literal starvation in the midst of plenty--situations found in many other countries.

A 1977 estimate placed over 13 million Tanzanians, about 80 percent of the population, in 7,684 *ujamaa* villages. Although there was maladministration and even mistreatment of people in carrying out the villagization program, it is clear that it had the support of most citizens.

> [It] remains true that 11 million people could not have been moved by force in Tanzania; we do not have the physical capacity for such a forced movement, any more than we have the desire for it (Nyerere, 1977: 42).

Critics like Hill, Shivji, and Cooper appear to have written off the experiment as a failure in that it is not ushering in an era of scientific socialism. Rather, a bureaucratic elite is entrenching itself as a national bourgeoisie. Other critics say that the economic and social costs have been too great and

Plate 3. Two old Kamba women shelling maize.

Plate 4. Kitemu wa Nguli and part of his family.

Plate 5. A woman reputed to be over 100 years old and several children in Ngelani, Machakos District, an area of considerable male emigration.

Plate 6. A causeway that has survived the rain from September 1961 to May 1962.

Plate 7. Beehives in a baobab tree (*Adansonia digitata*).

Plate 8. Part of the main field of Kitemu wa Nguli: *Acacia-Commiphora* bush with scattered *Euphorbia* forms the background.

Plate 9. Isaac Mwalonzi wa Ngea, one of the leaders in a 1938 protest against a government destocking scheme for Machakos District.

Plate 10. Nathaniel Venge of Kangundo helping to pick his coffee crop. He was aided under the Swynnerton Plan.

CHAPTER IV

THE IMPOVERISHMENT OF EAST AFRICAN DIET

Although East Africans, except along the Islamicised coast, do not have a "cuisine" in the sense that one finds a cuisine in West Africa, France, or China, the diet has traditionally consisted of a wide variety of foods, and variety continues to characterize the diet in some areas (Kreysler and Mndeme, 1975: 23). Nonetheless, nutritionists have noted over the years a considerable simplification and impoverishment in the diet of East Africans.

Despite regional differences, especially in areas where plantains are a staple, the standard fare in most East African farmers' homes is *ugali*, a stiff porridge made of steamed maize meal, although millet and sorghum grains and cassava flour are also used. It may contain other ingredients, or be consumed with a variety of relishes made of beans, meat, or boiled greens. The next most common food is *uji*, a gruel or soup, also usually made of maize and drunk from bowls. *Uji* is often the breakfast meal. Locally made beer is an important component of diet in many places, and is made from maize, sorghum, sprouted finger millet, bananas, or sugar cane. Mead beer and palm wine are also important in semi-arid and coastal areas, respectively. Tea bought from shops is also a common drink.

This diet may be contrasted with that of the present-day Sandawe, who live in the semi-arid zone of Tanzania north of Dodoma. As recently as the mid-1800s they were hunters and gatherers; since then they have become farmers and taken to keeping livestock. Their rich and varied diet can be taken as representative and perhaps even typical of the diets of the people of the semi-arid and sub-humid areas of East Africa in pre-colonial times. Newman (1975), who studied their diet in the mid-1960s, found that although *ugali* was the main preparation at 91 percent of the mid-day and evening meals, roasted potatoes, groundnuts, cowpeas, bambara nuts, and cassava were included at times, and, more significantly, the _ugali_ was accompanied over 45 percent of the time with a wide variety of relishes--made from boiled greens, mushrooms, gathered seeds, eggs, milk, fish, and over twelve kinds of meat, wild and domestic. The grain beer prepared from red sorghum or finger millet is also a valuable item in the diet of everyone, including children.

In addition, the Sandawe supplement their diet with at least 40 kinds of fruits, nuts, seeds, and berries collected in the bush, as well as several kinds of termites, larvae and caterpillars. These foods are usually taken in the course of herding, hunting, collecting firewood and travelling. Honey is a very important food for the Sandawe, and because of the way they set out and manage the hives, it is plentiful and available in all seasons. The Sandawe dry and store many sorts of harvested and gathered foods, and smoke fish to use during the dry season (May-November) when cows have ceased to give milk and there are no green leaves to gather for relishes. Finally, in times of food shortage the Sandawe rely on special famine foods which they gather. Newman listed ten famine foods and stated that although "these foods are bitter tasting and, due to their high fiber content, difficult to digest,...historically they have provided life-maintaining energy for many people who without them undoubtedly would have perished from starvation" (Newman, 1975: 37-38). Medical surveys among the Sandawe show them to be in comparatively good health. "Protein calorie deficiency of the kwashiorkor type is virtually non-existent indicating that not only are the resources for weaning available, but also that the mothers have adopted satisfactory weaning practices" (Newman, 1975: 38).

Nutritional surveys in many other parts of East Africa, however, reveal widespread nutritional and protein deficiencies, particularly in areas where children have little access to dairy and meat products.[1] A survey in the semi-arid zone in Kenya near

---

[1]See Latham, 1965; Amann, Belshaw and Stanfield, 1972; Rutishauser, 1974; and Sharman, 1970.

Machakos found that from 15 to 45 percent of the children aged 3 and under were less than 80 percent of the standard weight for their age, which is an indication of high vulnerability to measles, pneumonia, tuberculosis, and diarrhea (Wisner and Mbithi, 1974: 93; Wisner, 1977b). If Sandawe-like diets were the norm in earlier times, and the accounts of early observers show this to be the case, why has the diet of East Africans been impoverished? The changes in diet are of the following sort: (1) an increased reliance on maize and cassava and a diminution in the importance of several sorts of sorghum, finger millet, and bulrush millet. Heavy reliance on maize in the absence of compensating foods, exposes a population to the risks, both neurological and gastro-intestinal, of pellagra, a vitamin B (niacin) deficiency; (2) decreased reliance on bush foods, hunting, and the products of livestock; (3) acceptance of high-cost prestige foods, some of it imported from western Europe--tea, wheat flour, golden syrup, commercial beer and soft drinks, and canned food, although little of the budget of really poor farmers would be spent for such things; and (4) some changes in weaning practices and use of powdered milk formulae and bottle feeding of infants.[1]

Causes for these dietary changes are complex and numerous. Maize has displaced other grains because it generally gives greater yields, can be sold in the national market as well as consumed, resists bird damage better than open-pannicled grains, and governments have encouraged its use. Government also promoted French and haricot beans which displaced superior calcium-rich local beans (Taylor, 1974: 85). Population increases and governmental and settler land alienations reduced the amount of land

---

[1]Monckeberg in discussing the health implications of bottle feeding in underdeveloped countries stated: "[I]n a recent investigation done in a slum area population, it could be demonstrated that approximately 80 percent of the feeding bottles contained Shigella, Samonella, or E. coli when the mother was giving it to the child. This means diarrhea and gastro-intestinal disturbances. In the summer months, 30 percent of the calories of children living in such a population was lost due to diarrhea" (1977: 123-24).

available for stock grazing, and reduced the per capita availability of meat and dairy products for people in some semi-arid areas. The growth of a cash crop, such as cotton, has diverted both land and labor from food production. One typical change among East Africans has been to substitute cassava for a more nutritious staple, because cassava requires little work and this labor can be used for the cash crops. Two other causes for decline in the use of bush foods are that farmers, especially women, have less time to look for them and young people are ignorant about them. Tallantire found that "The increased pace of life no longer allows women to spend such long periods away from the home for the collection of leaves and fruits over considerable distances" (Tallantire, 1975: 244). The variety of collected and cultivated plants is more limited than it was in the past. Furthermore, knowledge of wild plants and game is decreasing, as young people, who hunt and gather less than did their parents, come to rely on a small array of domesticated plants and purchases of processed foods.

Food avoidances also have important dietary consequences for people in semi-arid areas. People of Islamic faith avoid pork. Fish, despite Sandawe liking for it, is widely avoided by people in semi-arid East Africa (Figure 10). Simoons, after reviewing the evidence of fish avoidance in Africa, advanced the hypothesis that it originated among pastoral people, who had a ready source of protein in their stock.

> Where fish avoidance occurs, moreover, it is poor individuals and groups who are most likely to consume fish, an act that confirms their lowly status. To me, it is a reasonable hypothesis that African fish avoidance (avoidance of fish in general rather than of particular species) had its origin in the scorn pastoralists commonly display for farmers and their ways (Simoons, 1974: 102).

Whatever its origin, avoidance of fish deprives large numbers of people of a valuable source of high quality protein.

Finally, a serious deficiency in the diet of East Africans results from inefficient storage and preparation of food.

FIGURE 10

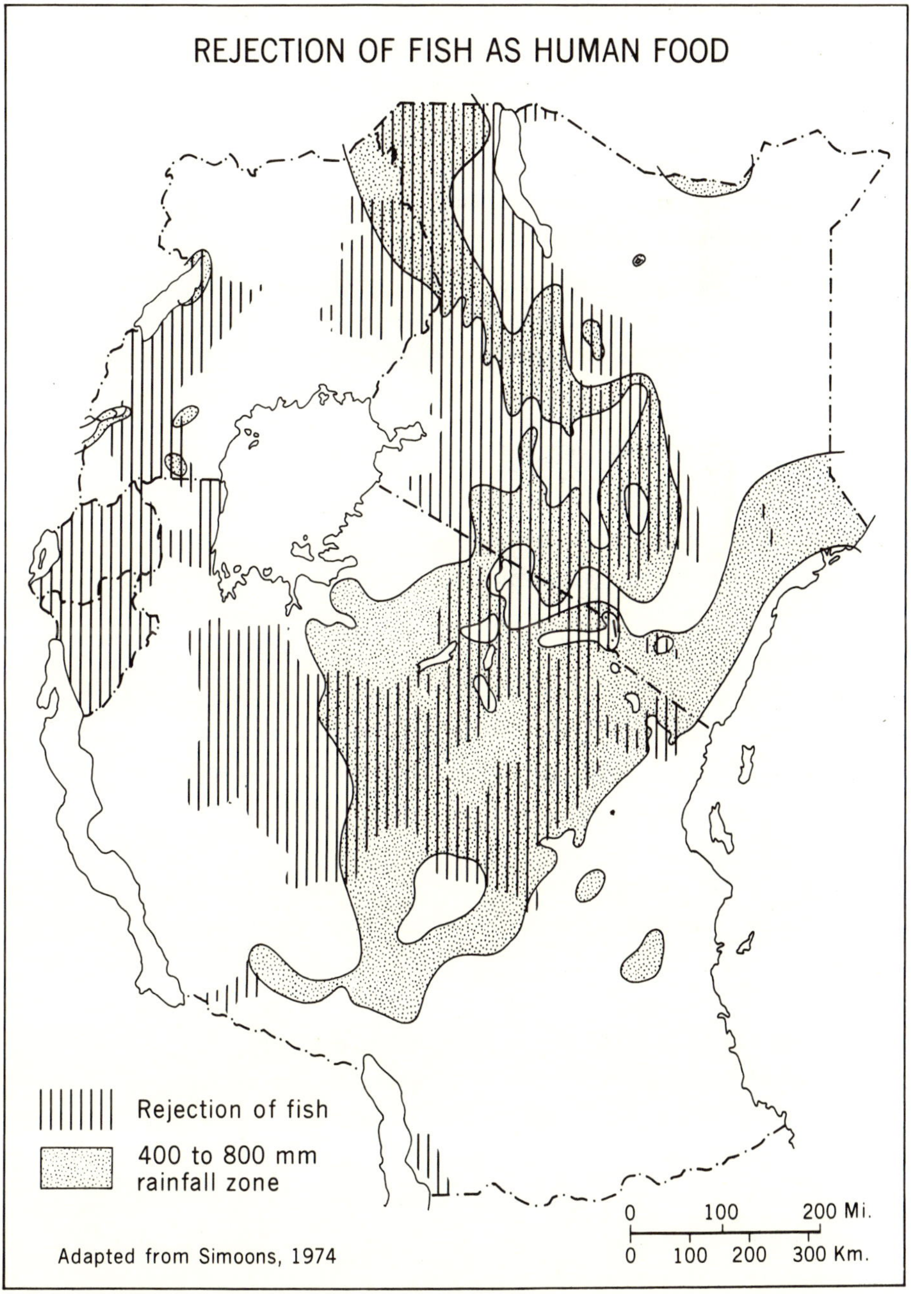
REJECTION OF FISH AS HUMAN FOOD
Rejection of fish
400 to 800 mm
rainfall zone
0 100 200 Mi.
0 100 200 300 Km.
Adapted from Simoons, 1974

Estimates vary on the amount of stored food lost to weevils, rats and mice, other insects and animal vermin, as well as rot, mold and bad handling. Losses of 25 to 50 percent are not uncommon. The Organization of African Unity estimated in a 1967 study that 30 percent of the grain stored in Africa is lost in storage alone (Owen, 1973: 98). Losses to agricultural vermin prior to harvest are also large, and too frequently total. An article in *Ceres* noted another cost with respect to food losses:

> [V]ery roughly [food losses] amount, according to the various cases, to between 30 and 50 percent of what has been grown (whether still growing or already stored). But the cost of replacement is two to three times greater because the food that must be imported to replace the lost crops is more elaborate and needs conditioning, and transport costs must be added (*Ceres*, 1976: 8).

During preparation, much of the food value may be lost. For example, in the Usambara Mountains of Tanzania, in the local preparation of maize to produce the flour (*unga*) from which *ugali* is made, extraordinary losses of protein, fat, amino acids, minerals and vitamins occur, as well as a significant diminution of carbohydrates (Table 12). Losses result at each step of the flour-making process: first the grain is pounded in a wooden mortar (husks and germ are separated and fed to poultry); then the pounded grain, called *pure*, is soaked for three or four days until it begins to ferment, at which point the water is drained off and the *pure* is washed three times with fresh water, which is decanted. The remaining *pure* is dried and pounded into a semolina-like flour, or *unga* (Schlage, 1968). As a result of this process the family, starting with 100 grams of maize, has lost 40 of the original 86 grams of dry matter. Protein loss is about 50 percent. Fat is one-sixth its original value. Minerals (ash) are reduced to less than one-tenth their original amount. The repeated leachings together with reduction of fat means that there are high losses of both water soluble and fat soluble vitamins (carotin, riboflavin, thiamine, and niacin), iron, and essential amino acids.

TABLE 12

NUTRIENT LOSSES IN LOCALLY PROCESSED MAIZE, USAMBARA, TANZANIA

	Maize	Pure	Unga
Total weight (grams)	100	77	72
Water	14	26.5	26
Total dry weight	86	50.5	46
Carbohydrate	70	44.5	41
Protein	8.5	4.4	4.2
Fat	3.8	0.7	0.6
Ash	1.2	0.2	0.1
Fibre	2.7	0.6	0.4

(After Schlage, 1968)

Schalge estimated in 1968 that the food lost through local hand milling (which accounted for roughly half of all milling in Tanzania) could provide the calorie requirements of over 900,000 adults, something which could be achieved by using small milling machines. A further point is that it takes a large amount of heavy work on the part of women to produce this nutritionally inferior result. Food storage and preparation are important topics for food policy analysis, and their improvement would do much to increase the quality and quantity of the diet of rural people.

Nutrition links directly to all other aspects of development. Its improvement is a practical concern of great importance. Scrimshaw (1966, quoted by Sharman, 1970: 77) summarized the many human costs of malnutrition:

> [T]he cost of malnutrition in less developed areas is exceedingly high; it includes the waste of resources in rearing infants who die before they can become useful citizens and the reduced working capacity of malnourished adults. From a quarter to a third or more of the children die before they reach school age, largely from infections that would not be fatal to a well-nourished child or from clinical malnutrition precipitated by a prior episode

> of acute infectious disease.... Nearly all children among the less privileged populations of underdeveloped countries show retarded growth and development at the time they reach school age; and although they are rarely seriously malnourished during school years, they do not make up for the deficit acquired during preschool years.

Recent evidence suggests that retardation in physical development in infancy is paralleled by impaired mental development that is probably permanent. This means that the future development of a country is compromised by serious malnutrition in young children. Moreover, attempts to provide adequate medical care are complicated and made more costly by both the greater amount and longer duration of illness of both malnourished children and adults.

CHAPTER V

STATE FOOD POLICY

State policy with regard to food production, pricing, and shortages has enormous consequences for the people of the semi-arid zone (Johnston, 1974). It is examined here first for the colonial era and then for the period since independence. Under British colonial rule in all three East African territories objectives of the administration were to monitor food production as it developed through the agricultural season, to provide food relief in areas of severe shortages when necessary to prevent starvation, and to see to it that people who lived in areas prone to food shortages grew "famine" crops, so that in bad years they would have something to fall back on. These various policies can be illustrated as follows:

Each month in each district (that is, in about 100 areas of East Africa, 30 in Kenya, 55 in Tanzania, and 16 in Uganda) the agricultural officer wrote a "monthly letter," a report describing the weather, the activities of farmers in different parts of the district, and the current food situation. An extract from Iringa dated February 26, 1962 is drawn from one such report:

> Weather - Overcast cloudy, general lack of sunshine. Heavy rain with days of light drizzle. Temperatures have been generally lower than usual for this time of year.
>
> Maize - Good crops cannot be anticipated. Much of the early planted maize is flowering before the maize has attained any size. In highter rainfall areas in particular the crop will be very small. Damage by stalk borer has been extensive through most of the southern part of the district.
>
> Beans - Yields will be average.

> Groundnuts - the very wet conditions have caused growth to be slow. Yields likely to be fair.
>
> Eleusine - Less satisfactory than normal.
>
> Potatoes - Increased acreage of both sweet and European varieties being reported for all areas as a result of the poor growth of grain crops....
>
> Food situation - Tanganyika Millers report current stores of 20,000 bags [1,818 metric tons] of maize. Resale setup in hand for Malangali, Mlolo, Idodi, Mazombe, and Pawaga Chiefdoms. Extensive flooding in Pawaga now. River is over two miles wide.
> W. C. Mathewson, 1962.

These reports were collated at regional and national levels and estimates made of potential harvest and whether or not food shortages necessitating relief would develop during the next few months. Sufficient grain stocks were kept in hand to meet local short-lived food relief requirements, but very serious shortages required imports from other East African territories or overseas. The monthly reports were supplemented by other estimating techniques, for example, field sampling of crops, granary inspection, visits to markets, and interviews with traders.

In Uganda, the local chiefs were required to provide "Monthly Planting Returns" wherein they stated the number of plots of each crop which had been started within their jurisdiction that month. These were then multiplied by an average plot size to estimate how much land in each crop had been planted. Average plot sizes for each district had been determined earlier by the Department of Agriculture and the East African Statistical Department, which had measured for the most important crops 100 randomly chosen plots in each district of the country (East African Statistical Department, 1959; Uganda Protectorate, 1961). These acreages and plot counts, together with the monthly letters, provided a numerical basis for estimating crop prospects.

Fuggles-Couchman described the three types of measures taken in Tanganyika to combat famine:

> [T]he maintenance of adequate reserves of root crops in the ground; the storage of surplus grain in those districts most likely to suffer from food shortages; and restrictions on sale and movement of staple crops in districts when a food shortage is threatened (1964: 38).

The planting of cassava and sweet potatoes was made mandatory for people in areas prone to food shortages and crop failures. In some cases communal planting of cassava was encouraged. Native Authorities "were encouraged to build small grain-storage silos, in which all cultivators had to deposit specified quantities of harvested grain available for distribution in times of food shortage" (Fuggles-Couchman, 1964: 89). These planting and storage requirements and restrictions on inter-district trade were exceedingly unpopular, and together with compulsory terrace construction and destocking schemes (which were undertaken because of widespread soil erosion and gullying) they became an issue around which African political mobilization and protest emerged (Young and Fosbrooke, 1960; Fuggles-Couchman, 1964; Ehrlich, 1974; Munro, 1975).

Very little has been written on food relief which reveals contemporary governmental attitudes, but they may be deduced from governmental action and inaction. Food relief for crop failure is intended as emergency relief and classed with other natural disasters such as earthquakes, floods, or typhoons. Relief is for extraordinary circumstances and governments dislike the idea of its becoming institutionalized as a form of welfare for areas that experience chronic food shortfalls. The evidence of Machakos District, Kenya, in the decades just prior to independence, suggests that farmers may come to depend on food relief, and to treat it as just one other feature of the agricultural system, in this instance an insurance mechanism provided by government. Indeed, relief efforts were even abused at times by some who re-exported famine relief maize illegally, or used relief maize to pay part of their bride-price. Table 13 shows the food situation and famine relief in Machakos District from 1939 to 1962. In this 24-year period maize was imported in 19 years and in two others

TABLE 13

MAIZE IMPORTS AND EXPORTS AND FAMINE RELIEF
MACHAKOS DISTRICT, KENYA, 1939-1962

Date		Food Situation (+ grain exported, - grain imported) (Grain amounts given in metric tons)	Rain-fall in mm*
1939	±	Large quantities imported, some exported	529
1940		Exports prohibited	791
1941		Severe shortage early in year	1,073
1942	+	18,182	927
1943	-	10,727 Not all maize, some cassava flour, other cereals and pulses imported. Very bad year.	523
1944	-	21,340 Severe drought, locusts plentiful.	867
1945	-	17,091	483
1946	-	22,544	688
1947	±	2,869 imported Jan.-March, then large surplus crop, some exported	1,242
1948	+	Small exports early in year, then exports prohibited.	961
1949	-	8,348 Both rains failed.	589
1950	-	21,973 Maize, some wheat and beans imported. Severe drought.	651
1951	-	6,769 Importation stopped in Aug., good long rains harvest.	1,584
1952	-	Only township required food imports.	788
1953	-	10,793	758
1954	-	6,450	929
1955	-	7,263	936
1956	±	5,156 imported. About 1,818 tons (20,000 bags) exported illegally and to Maize Control	540
1957	±	1,585 imported for township, 6,306 exported to Maize Control	1,042
1958		Small surplus	1,022
1959	-	4,545	672
1960	-	15,455	796
1961	-	27,273	1,524
1962	-	Over 27,273 (300,000 bags)	806

* Recorded at District Commissioner's boma.

Sources: Machakos District Gazetter, n.d., East African Meterorological Department, and field notes.

exports were prohibited. In only three years was there a significant surplus. This record of food shortages occurred despite many years when rains were average or good.

Edgar Winans, who inquired into food policy in East African countries, found the following:

> Neither Kenya nor Tanzania has issued a White Paper or announced food shortage policy goals through such documents as the Development Plans or annual pricing announcements.
>
> The governments of Kenya and Tanzania have never made an official public statement on famine policy. However, both governments do produce yearly internal documents in connection with pricing decisions which forecast expected yields of the principal cereal crops and which contain recommendations for retention of emergency reserves. These documents are regarded as highly confidential since they determine the level at which exports will be allowed. They are therefore unobtainable outside of Government. In Kenya there is a line item in the yearly budget for the Office of the President for famine relief. In some instances this has been shown as a token vote of one Kenya pound simply as a means of retaining the item for use if emergency drought requires it.... Certain unofficial statements have been made by high ranking government officials in both Kenya and Tanzania concerning government intention to maintain strategic grain reserves. It is the current intention of the government of Tanzania to build up a 2 to 2-1/2 year reserve. Kenya Government seems to plan a reserve of approximately one year. In both cases the policy is specific to the problem of recurrent drought and chronic food deficits in specific districts (Winans, 1977).

In Tanzania, the National Agricultural Products Board (NAPB) has been responsible for collecting crop data and making recommendations regarding pricing, storage, and imports and exports of food.[1] Crops offered for sale are supposed to be marketed through the NAPB at prices fixed for the season during the preplanting period at a time when no one can know whether the rains that year will be good or bad. Thus, if production is

---

[1]More recently the NAPB has become the Marketing Development Bureau of the Ministry of Agriculture.

greater the NAPB is forced to buy at the subsidized price and either place the surplus in storage, or export it at a loss, since the world prices for grains such as maize and rice are commonly lower than the NAPB prices to farmers. If the harvest is poor, demand increases, and in any area with a local surplus there will be a black market price in excess of the NAPB price, and thus farmers will refuse to sell to NAPB. A black market movement of grains often develops between Kenya and Tanzania when price differences between the two countries are great. That the NAPB had not done a good job of estimating production and setting prices is attested to by the statement that "in six out of the eight years since the NAPB was set up it has both imported and exported maize in the same years, and taken losses on both sets of transactions" (The Standard, 26 June 1971).

Serious disruptions in food supplies were sometimes traceable in part to hoarding by merchants, who hoped by withholding commodities to drive up prices and thus sell later for greater profits. During the first decade of independence in Tanzania, newspapers would from time to time carry exposés of traders (usually Asian and Arab shopkeepers were featured) who had large inventories of scarce commodities, such as flour, sugar, rice, or kerosene.

On food policy and agriculture, Nyerere stated in 1977:

> One other tradition to which we must return is the storage of grains in the villages. Traditional storage methods could often be made very effective if minor changes were made to keep out vermin. And village food stocks would provide additional security for people against future disasters, including a breakdown of communications, and would save the peasants' cash income for other things. It will, however, still be necessary for us to build up national stores of food in different parts of the country--about which we have been remiss in the past (Nyerere, 1977: 22).

In April, 1977 a National Food Security Programme was launched by the Tanzanian Ministry of Agriculture. It involved the storage of 100,000 tons (equal to 11 million 200-lb. bags) of grain, mostly maize, rice, sorghum and wheat, the equivalent of about three months' nutritional requirements during a crop failure. Although buildup of the reserve (from a current capacity of 60,000 tons) was to be based on imports from overseas, it was hoped that by 1980 the country would have achieved self-sufficiency (*Daily News*, 1977:1).

The food imports of 1974 and 1975 (Figure 1) were unusual, but they show that Tanzania:

> is still economically dependent upon the vagaries of the weather, and upon economic and political decisions taken by other people without our participation or consent. And this is not a reciprocal situation, Tanzania is still a dependent nation, not an interdependent one (Nyerere, 1977: 2).

Nonetheless, Tanzanians have a greater control over their lives and resources than formerly.

> It is the people as a whole who determine the kind of society we shall create, and not foreign capitalists, or even local capitalists. And thus we have been able to use that control for our own benefit. Thus, for example, we were able to marshall all our foreign exchange resources to fight against the serious famines of 1973-75, so that no-one in this country died of hunger or was even seriously affected by it in the way which used to happen in the past.... We used up all our foreign exchange reserves in fighting the drought and meeting the increased costs of essential fuel and development goods (Nyerere, 1977: 49-50).

Events in recent decades in Tanzania have taken a very different course in comparison with what has occurred in Kenya. Tanzanian institutions and governmental practices place the people of semi-arid areas on a more nearly equal footing with those living in higher potential and urban areas.

CHAPTER VI

APPROPRIATE SCIENCE AND AID IN SEMI-ARID EAST AFRICA[1]

We return finally to a question posed at the outset: What can those who live in developed western industrial countries offer in the way of agricultural technology to people in underdeveloped

---

[1]Edgar Winans, who provided thoughtful comments on issues raised in this study, noted the rising costs of larger social orders, and the relations places and groups have to larger social orders.

> The marginal zones of such places as Eastern Africa cannot yield a return sufficient to carry the costs of large scale social systems, yet at the same time, and because of the small scale nature of the organization they will support, they become highly vulnerable to economic exploitation. All that need be true for such exploitation to occur is that there be a cline in costs. These marginal zones need not yield very much, they just have to yield it at low cost. Thus they may be exploited for their labor supply because that labor supply is insufficiently organized to resist exploitation. They may be exploited for their food in a good year because the internal marketing mechanisms are too small in scale to bulk and store, while the marketing system of adjoining, better endowed regions is well set up to siphon off the food at very low prices. These and other mechanisms operate to make semi-arid environments a part of large scale social systems...[and] I don't see how the marginal areas can be pulled out of the larger social systems in which they have been incorporated....
>
> [Under] current circumstances of the expansion of the scale of social systems, people occupying semi-arid regions are at a chronic disadvantage in the terms of trade they "enjoy" with other regions. This is one instance of a world-wide process (rural areas have a negative balance on trade with cities, tropical areas suffer in trade with temperate zones, agrarian countries with industrial countries, and aside from Opec, nations with extractive industries are at a disadvantage vis a vis final producers). The world as a whole faces a crisis of organization. We must find a means of damping the forces which cause imbalances in flows of resources (people, materials, food, whatever). We have gone from small scale systems that operated by redistribution to large scale systems that operate by aggregating resources and focusing their reinvestment. This allows growth while the former state allowed equity. How do we get both? [A key question remains to be answered:] how to arrange the social order so as to allow the utilization of these marginal zones without exploitation of them or the people who live in them (Winans, 1978).

countries? In amendments to the Foreign Assistance Act in 1975, the U.S. Congress issued a mandate to the Agency for International Development (AID) for "a broadly participatory agricultural production strategy for developing countries." The concern is to help underdeveloped countries to: (1) increase their capacity to expand and distribute food supplies to alleviate hunger, and (2) increase participation of poor people in the process and benefits of development" (A.I.D., 1977: 1).

> Section 103 on 'Food and Nutrition' of the Foreign Assistance Act (1975) states these objectives quite explicitly. Section 103(a) reads, in part, '...to alleviate starvation, hunger, and malnutrition, and to provide basic services to poor people, enhancing their capacity for self-help...' Section 103(c) states that assistance under the section '...shall be used primarily for activities which are specifically designed to increase the productivity and income of the rural poor...' (A.I.D., 1977: 1).

An A.I.D. paper responding to the Congressional mandate recommends a program in terms of five functional categories, which are in rank order: (1) Land Reform and Local Institutions, (2) Planning and Policy Analysis, (3) Research, Education, and Extension, (4) Rural Infrastructure, and (5) Marketing and Storage, Input Supply, Rural Industry and Credit.

> The ranking is based on our analysis of agricultural development experience which suggests that, if an activity in one category is to address both production and equity objectives successfully, the functions served by activities in higher ranking categories should be satisfactorily addressed.... For example, a decision to initiate or strengthen a small farmer credit program should normally be based on findings that the following are adequately being addressed: a reasonably equitable distribution of land-holdings and effective participatory institutions of governance at the local level, appropriate economic policies (e.g., those affecting input and output prices, including interest rates), accessible and profitable technology, and related supporting rural infrastructure and marketing services (A.I.D., 1977: 19-20).

To ask industrial mid-latitude agricultural science to aid agricultural development in semi-arid East Africa, or elsewhere in the underdeveloped world, brings two food technologies face to face, each embedded in a different social setting (Knight and Wilcox, 1976). I have already described lineage agriculture. By contrast, U.S. agriculture can serve as an example of industrial agriculture (Figure 11). Key elements are: capital intensity, technological inputs, high energy use (relying on electricity and fossil fuels), concentration of economic power because of the vertical structure of ownership, production, processing and distribution, and a quest for profits. It is a system that works in the aggregate when there is drought because it parts are interlinked, and production can be moved to any consuming area, farmers can purchase insurance against crop failure, and government provides crop subsidies, low interest loans and disaster insurance and relief. If we take a figure representing the structure of the U.S. food system and adapt it to semi-arid East Africa, we have to start by eliminating items. We end up with the more simple appearing system discussed earlier (Figure 8, page 33). Its key elements are low technological inputs, reliance on sun, fire, and human (possibly animal) labor for energy, disaggregation of economic power or concentration of it in certain groups: older men, women if they control production and marketing, royal clan members. In any event, agricultural work is not concentrated in the society. It is a system that works (or used to work) when there is drought because of the large set of social and environmental practices which ensure success of crops, protect livestock, or share out hardship among a larger kindred.

The two technologies--modern industrial agriculture and lineage agriculture--are incompatible at many points. As farmers adopt parts of the modern technology, they have to give up practices that were functional to the family lineage technology. The land-population-migrant labor squeeze in the semi-arid areas sometimes forces people to give up lineage mode practices, but prevents them from using modern practices, which increases their environment's vulnerability to drought.

FIGURE 11

THE STRUCTURE OF THE U.S. FOOD SYSTEM

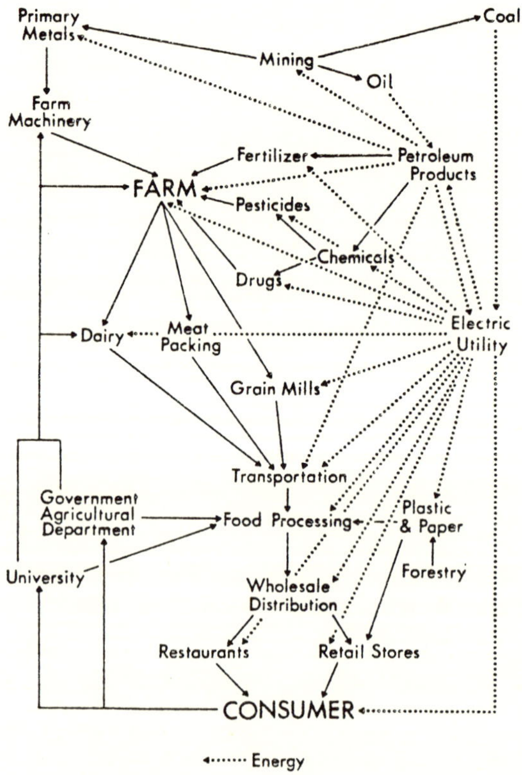

Source: Knight and Wilcox, 1976.

An example of technical inputs requiring new behavior and the setting aside of former practice is provided by Katumani maize (Figure 12). Katumani outyields Local Farmer's maize if the rainfall is less than 340 mm.[1] Beyond that Local Farmer's maize does much better, a point of great unhappiness among farmers when rains are good. Katumani has to be monocropped to use the moisture to advantage and to yield well. It requires a lot of care and special effort on the part of the farmer, as well as special inputs (Table 14). Many farmers are in no position to buy seed, insecticides, and fertilizers, yet if they do not buy all three of these together and follow good practices, the potential yield increase is not realized.

It is not at all clear that people in underdeveloped countries should or could achieve an industrial agriculture similar to that which has developed in the U.S. American farmers and agribusiness corporations are interested almost exclusively in the exchange value (profitability) of what they grow. American agricultural technology is designed to meet requirements of efficiency and profitability. It seeks ways to be labor saving; it is energy spending. East African farmers are interested in both use and sale of their crops. Between 80 and 90 percent of East Africans are engaged in farming, herding, fishing, or some combination of these pursuits. There is no prospect that a large proportion of these people will be drawn out of rural areas to take jobs in industry and commerce, particularly in view of projected growth of the population--to 90 million by the end of the century. Agriculture in particular will continue to be a way of life for most East Africans. Those attempting to find ways to improve its productivity and reliability need to keep that fact in mind.

In this paper it has seemed more important to analyse environmental and socio-political aspects of food production in

---

[1]Recent developments with Katumani seed have increased its ability to match yields with local maize (Local Machakos White) when rains are greater, although a difference in yield apparently persists.

FIGURE 12

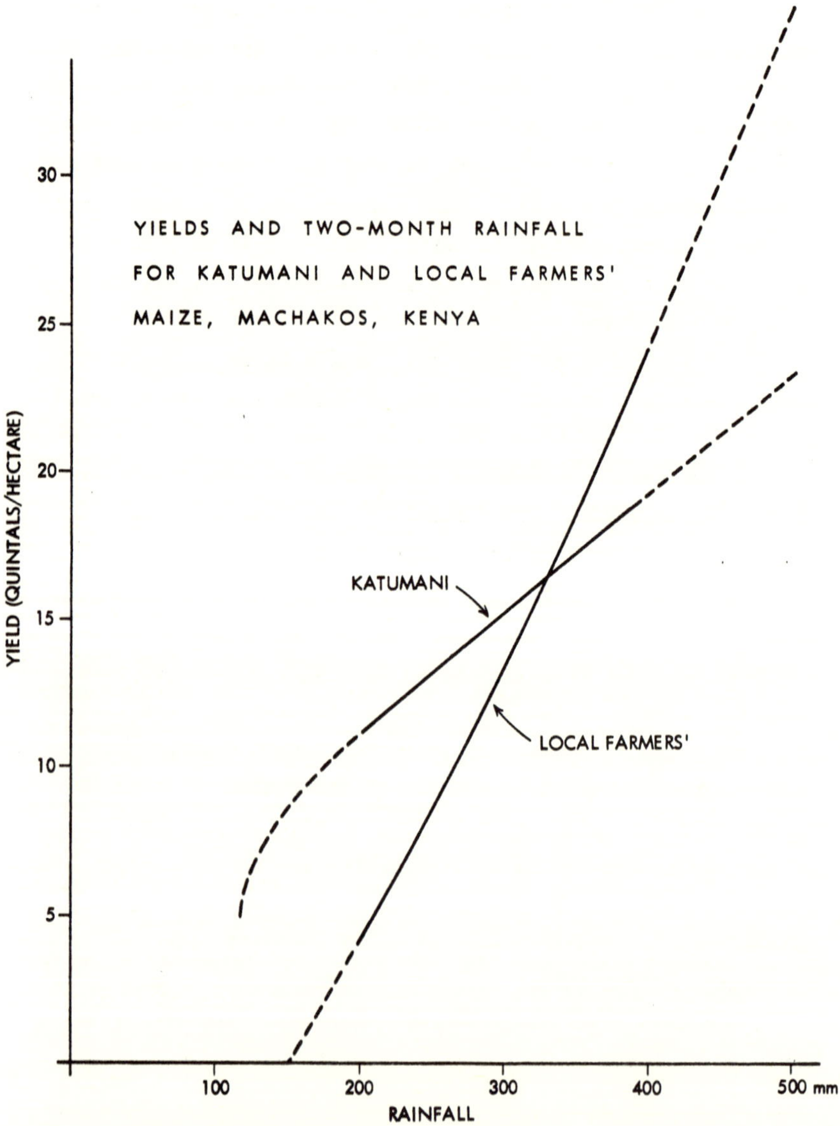

SOURCE: ANNUAL REPORT, MACHAKOS RESEARCH STATION, 1968

TABLE 14

PACKAGE OF HUSBANDRY PRACTICES REQUIRED TO MAKE KATUMANI YIELD EFFECTIVELY

(1) Preparation of a Ridged Seedbed
(2) Planting Before the Rain Begins
(3) Row Planting
(4) Spacing 3' by 1' (90 cm by 30 cm)
(5) Seed Dressing Against Fungal Infestation
(6) Sowing Two Seeds Per Hole
(7) Early Thinning
(8) Early Weeding
(9) At Least Three Weedings
(10) Fertilizer
(11) Dust Against Stalk Borer
(12) Monocropping Only

Source: Wisner, 1977a.

semi-arid areas than to review efforts currently underway whose purpose is to change the nature of rural production and employment. Some recent research and extension-like programs exhibit an awareness of lineage agriculture as a system-with-a-history. One example from Kenya is that of CIMMYT (Centro Internacional de Mejoramiento de Maiz y Trigo) Eastern African Economics Programme. Although CIMMYT's charge is to improve "maize and wheat by breeding better genetic material," its economics program seeks to establish cooperative inter-disciplinary research so that experimental programs will reflect both natural and economic circumstances and produce technology appropriate to small farmers (Collinson, n.d.).

Perhaps the most succinct exposition of problems of bridging the chasm between poor rural people and those concerned with the design and transfer of technology is provided by Mbithi (1977). He states that "change agents" and the farm householder have uninformed, confused, contradictory views of one another. He comments on several social psychological aspects: dependence, alienation, resource immobility, and suppression of indigenous adaptation techniques.

> [T]he change agent in East Africa is a member of the educated elite, who became alienated from his illiterate parents when he went to school at the impressionable age of between six and eight, spent at least 15 years 'learning' western value systems, and life styles. He shares little with his rural folk and has no empathy with their thought processes. He may be arrogant or paternalistic but he does not recognize them as independent managers of their holdings....
>
> [T]he rural farmer has, for the past 70 years ... been told repeatedly that he is backward, heathen, ignorant, illiterate and headed straight to "hell". ...[T]his orientation, [exalting[the supremacy of western technology, western value systems and western religions, saps self [assertiveness] and self identity from the humble rural farmer, and even makes him apologise for daring to see things differently. This approach generates a school-boy mentality among rural adults who are very willing to suspend their world view, abandon it and embrace that of the expert, who may have just recently scraped through an agricultural exam based on a narrowly designed curriculum (Mbithi, 1977:30).[1]

Mbithi's conclusion supports those of this paper. The existing indigenous agricultural systems and the social setting of which they are a part need to be taken seriously (Greenland, 1975). If they were properly understood by western scientiists, various means of improving their productivity and reliability without destroying the social milieu which surrounds them could be found. In moments of fantasy I imagine placing all western-trained aid experts individually and permanently in rural areas, cut off from the support of laboratories, computers, libraries, and peers. The prospect of making an agricultural living side by side with their new neighbors would "wonderfully concentrate the mind." I would predict two useful outcomes: (1) a new respect for and understanding of indigenous systems of production, and (2) the discovery of appropriate and useful ways to improve rural livelihoods.

A further conclusion concerns the development courses being

---

[1]My own impression has been that many poor people in semi-arid areas are not apathetic or defeated; they simply have less power and access to credit and new technology than their neighbors in high potential areas.

followed in East Africa. There are no easy solutions to livelihood difficulties for people in semi-arid areas. Limitations on human and livestock numbers living in semi-arid areas are no doubt desirable from an ecological standpoint, that is, a systematic balance between rate of use and rate of recovery from use, given the wide climatic variation characteristics of semi-arid areas. Control of these numbers would necessitate the provision of attractive alternatives for those who would resettle, or change their type of employment--and it is here that political, social, ethnic, and even psychological issues arise. In Kenya's case, only if the government treats the development of the semi-arid areas in the context of the resource development of the whole country can any long-term improvement of the life of poor people in the semi-arid areas be achieved. In Tanzania's case the most difficult task appears to lie with the government--navigation along a course which on the one hand does not (through over-regulation and bureaucratic inefficiency) stifle efforts of individual farm families to increase their productivity, and on the other hand prevents the emergence of a strongly differentiated class-based society.

To review two sobering facts of life: (1) the mid-latitudes will probably always be able to outproduce the tropics in grains by whatever measure, and accordingly will be able to control world prices of the main food grains; and (2) farmers in tropical semi-arid areas will always be in a disadvantageous position with repect to farmers in wetter, higher potential areas. The local advantages people in semi-arid areas have are: (1) a long growing season (even if yields are comparatively less), with the possibility of taking two and even three crops per annum from irrigated fields (with the labor implications that two and three seasons of work carry, if there are irrigable fields), and (2) the fact that they live on their own farms and have the opportunity to be self-reliant. Agriculture, particularly at national levels, should seek to be relatively self-sufficient, so as to dismantle the structural dependency of the colonial era which left large segments of the population in a situation wherein they grew products they could not eat and consumed food which they did not grow, but had to import from overseas. By reorienting internal production to meet internal needs, dependency on world commodity markets and price fluctuations can be reduced and a reliable and adequate indigenous food supply assured.[1]

The possibilities of improving a western science understanding of the environment and life of people in the semi-arid zone seem endless. Study of diet, storage technology,[2] food preparation, the agronomy of local food staples, child development, and disease should have high priority, as should research on indigenous agricultural systems in relation to agro-climates and soils. In preparing this study I read through microfilms of two newspapers: The Standard (Nairobi) and The Daily News (Dar es Salaam). From news stories in these papers it is clear that many of the food problems reported on are urban rather than rural. Their causes may be traced to poor administration and planning, bureaucratic red tape, unpopular pricing policies, hoarding, black-marketeering, breakdowns or disruptions in transport, and poor quality in manufacturing, handling, storing or marketing food. These matters, also, are worthy of serious research. Multidisciplinary studies of rural areas which incorporate a kind of ethnoagronomy and ethnoeconomics would be useful. A thorough understanding is needed of the political economy of livelihood in semi-arid areas, of the ways people there are marginalized politically and economically as well as environmentally with respect to high potential areas and within the larger system of world trade (Emmanuel, 1972). Studies are needed of ways rural people can increase control of their land and labor power, and collect the benefits of their industry. Methods developed by Paulo Freire among poor farmers in northeastern Brazil

---

[1]To argue for self-sufficiency is not to argue against trade and the specialization of production, either on national or world markets, where participation in these markets can be done on the basis of equality.

[2]See Anon (1977) for an interesting self-help approach to problems of food storage in which the local people took the lead in designing the study, carrying it out, and evaluating and implementing the findings.

could be explored in the context of self-help, self-reliance and *harambee* in East Africa (Freire, 1970, 1974). Finally, much will depend on the will of government to represent and act on behalf of all its people. If the contrasted policies of Kenya and Tanzania continue to operate over the next 10 to 20 years, we will probably be provided with definitive answers on state policy and life for people in semi-arid areas in East Africa.

REFERENCES CITED

Abelson, P. H. (ed.), 1975. Food: Politics, Economics, Nutrition and Research. Washington, D.C.: American Association for the Advancement of Science.

Acland, J. D. 1971. East African Crops. London: Longmans Group Ltd.

Agency for International Development, 1976. Research Literature for Development, Vol. 1, Food Production and Nutrition, December.

________, 1977. A.I.D. Agricultural Development Policy Paper, Draft No. 4, PPC/PDA, May.

Allan, W. 1965. The African Husbandman. Edinburgh: Oliver and Boyd, 483 pp.

Amann, V. F., D. G. R. Belshaw and J. P. Stanfield (eds.), 1972. Nutrition and Food in an African Economy. 2 vols., Kampala, Uganda: Makerere University.

Anon, 1977. Appropriate Technology for Grain Storage. Report of a Pilot Project, Community Development Trust Fund of Tanzania in Collaboration with Institute of Adult Education, Economic Development Bureau.

Awiti, A. 1972. Economic Differentiation in Ismani, Iringa Region: A Critical Assessment of Peasants Response to the Ujamaa Vijijini Programme, Economic Research Bureau, University of Dar es Salaam, mimeographed seminar paper.

Aylward, F., and M. Jul, 1975. Protein and Nutrition Policy in Low-Income Countries. New York: Wiley.

Berg, A., N. S. Scrimshaw, and D. L. Call (eds.), 1973. Nutrition, National Development and Planning. Cambridge: MIT Press.

Blue, R. N., and J. H. Weaver, 1977. A Critical Assessment of the Tanzanian Model of Development. Agricultural Development Council Reprint No. 30, July.

Bohdal, M., N. E. Gibbs, and W. K. Simmons, 1968. Nutrition Survey and Campaign Against Malnutrition in Kenya, 1964-1968. Nairobi: WHO/FAO/UNICEF.

Brooke, C. 1967a. "The heritage of famine in central Tanzania," Tanganyika Notes and Records, 67 (June), 15-22.

________, 1967b. "Types of food shortages in Tanzania," Geographical Review, 57, (July), 333-357.

Burki, S. J., and T. J. Goering, 1977. "Food problems of the low income countries," Finance and Development, 14 (June), 14-18.

Campbell, H., 1975. "Socialism in Tanzania: A case study," The Black Scholar, 6, 41-51.

Ceres, 1976. "New Sahel threats," 9 (May-June), 8.

Chang, J., 1968. "The agricultural potential of the humid tropics," Geographical Review, 53 (July), 333-361.

Chayanov, A. V., 1966. The Theory of Peasant Economy. Homewood, Ill.: The American Economic Association.

Claeson, C.-F., 1974. "Interregional population movement and cumulative demographic disparity," Geografiska Annaler, 56, Ser. B, 105-120.

Cocheme, J., and G. Zazzara, 1969. "A contribution to the forecasting of the early rains at Nairobi," East African Agricultural and Forestry Journal, 34 (January), 312-314.

Collinson, M. P., n.d., The Economics Programme of the international Maize and Wheat Improvement Centre - East African Region. Nairobi: CIMMYT.

Daily News., 1977. "Food security programme to be launched." 16 April, 1.

Darrah, L. L., and L. H. Penny, 1974. "Altitude and environmental responses of entries in the 1970-71 East African maize variety trials," East African Agricultural and Forestry Journal, 40 (July), 77-88.

Davis, J. T., 1977. "Development of the small-farm sector in Kenya, 1954-72." Canadian Geographer, 21 (Spring), 32-58.

DeMarco, S., and S. Sechler, 1975. The Fields Have Turned Brown - Four Essays on World Hunger. Washington, D.C.: Agribusiness Accountability Project.

Deshler, W., 1960. "Livestock trypanosomiasis and human settlement in northeastern Uganda," Geographical Review, 50 (October), 541-554.

deSouza, A. R., and P. W. Porter, 1974. The Underdevelopment and Modernization of the Third World. Resource Paper No. 28, Commission on College Geography, Association of American Geographers, Washington, D. C.

deWilde, J. (ed.), 1967. Experiences with Agricultural Development in Tropical Africa. 2 vols. Baltimore: The Johns Hopkins Press.

Dowker, D. M., 1971. "A note on the reduction in yield of Taboran maize by late planting," East African Agricultural and Forestry Journal, 30, 33-34.

Dyson-Hudson, R., and N. Dyson-Hudson, 1969. "Subsistence herding in Uganda," Scientific American, 220 (February), 76-89.

East African Statistical Department, Uganda Unit, 1959. Investigations into Acreage Statistics, Uganda Protectorate, September.

Egero, B., and R. A. Henin (eds.), 1973. The Population of Tanzania, Census Volume 6, Dar es Salaam: Jointly published by BRALUP and Bureau of Statistics.

Ehrlich, C., 1975. "Economic and social development before independence," in B. A. Ogot (ed.), Zamani. Nairobi: East African Publishing House Ltd. and Longman Kenya Ltd., 335-345.

Emmanuel, A., 1972. Unequal Exchange. New York: Monthly Review Press.

Freire, P., 1970. Pedagogy of the Oppressed. New York: The Seabury Press.

________, 1974. Education for Critical Consciousness. New York: The Seabury Press.

Fuggles-Couchman, N. R., 1964. Agricultural Change in Tanganyika: 1945-1960. Stanford: Food Research Institute.

Geertz, C., 1963. Agricultural Involution: The Process of Ecological Change in Indonesia. Berkeley and Los Angeles: University of California Press.

Goldschmidt, W., 1966. Comparative Functionalism. Berkeley and Los Angeles: University of California Press.

________, 1976. Culture and Behavior of the Sebei. Berkeley and Los Angeles: University of California Press.

Greenland, D. J., 1975. "Bringing the Green Revolution to the shifting cultivator," Science, 190 (28 November), 841-844.

Gulliver, P., 1955. The Family Herds. London: Routledge and Kegan Paul.

Hare, F. K., 1976. Climate and Desertification. Toronto: University of Toronto.

Hill, F., 1975. "Ujamaa: African socialist productionism in Tanzania," in H. Desfosses and J. Levesque (eds.), Socialism in the Third World. New York: Praeger Publishers, 216-251.

Huizer, G., 1978. "The ujamaa village program in Tanzania: New forms of rural development," Studies in Comparative International Development, 8 (Summer), 183-207.

Huxley, E., 1935. White Man's Country. London: Chatto and Windus, 2 vols.

Iliffe, J., 1971. Agricultural Change in Modern Tanganyika. Nairobi: East African Publishing House.

International Labour Office, 1972. Employment, Incomes and Equality. Geneva: I.L.O.

Jackson, I. J., 1969. "Tropical rainfall variations over a small area," Journal of Hydrology, 8, 99-110.

Johnston, B. F., 1974. Objectives and Scope of a Food and Nutrition Policy. University of Nairobi: Institute for Development Studies, Discussion Paper No. 210.

Karani, H., 1965. Pricing and Marketing of Maize in Kenya. University of Nairobi: Institute for Development Studies, Discussion Paper No. 19.

Kenya Population Census, 1962, Vol. 1, 1964. Nairobi: Directorate of Economic Planning, Ministry of Finance and Economic Planning.

Kjekshus, H., 1977a. Ecology, Control and Economic Development in East African History. London: Heinemann.

________, 1977b. "The Tanzanian villagization policy: Implementational lessons and ecological dimensions," Canadian Journal of African Studies, 11, 269-282.

Knight, C. G., and R. P. Wilcox, 1976. Triumph or Triage? The World Food Problem in Geographical Perspective, Resource Paper No. 75-3, Commission on College Geography, Association of American Geographers.

Kocher, J. E. (ed.), 1975. Social and Economic Development and Population Change in Tanzania, Research Paper No. 36, Bureau of Resource Assessment and Land Use Planning, University of Dar es Salaam, Tanzania.

________, 1977. A Bibliography on Rural Development in Tanzania, Development Discussion Paper No. 30, Harvard Institute for International Development.

Kraut, H., and H. D. Cremer (eds.), 1969. Investigations into Health and Nutrition in Africa, Afrika-Studien Nr. 42. München: Ifo-Institut für Wirtschaftsforschung.

Kreysler, J., and M. Mndeme, 1975. "The nutritional status of preschool village children in Tanzania: Observations in Lushoto District," Ecology of Food and Nutrition, 4, 15-26.

Latham, M. C., 1965. Human Nutrition in Tropical Africa. Rome: F.A.O.

________, 1967. "Nutritional studies in Tanzania (Tanganyika)" World Review of Nutrition and Dietetics, 7, 37-71.

Leakey, L. S. B., 1934, Science and the African, reedited by C. L. A. Leakey, 1970, mimeographed.

Legum, C. (ed.), 1977. Africa Contemporary Record, Annual Survey and Documents, 1976-77. London: Rex Collings, Entries on Kenya, pp. B218-250, and Tanzania, pp. B339-371.

Lele, U., 1975. The Design of Rural Development: Lessons from Africa. Baltimore: The Johns Hopkins Press.

Leys, C., 1974. Underdevelopment in Kenya: The Political Economy of Neocolonialism. Berkeley and Los Angeles: University of California Press.

Lofchie, M. F., 1976. "Agrarian socialism in the third world," Comparative Politics, 8 (April), 479-499.

Lowie, R. H., 1937. The History of Ethnological Theory. New York: Farrar and Rinehart, Inc.

Machakos District Gazetteer, n. d. (c. 1962).

Maro, P. S., 1974. Population and Land Resources in Northern Tanzania: The Dynamics of Change 1920-1970. University of Minnesota, unpublished Ph.D. dissertation.

Mascarenhas, A. C., 1971. "Agricultural vermin in Tanzania," in S. H. Ominde (ed.), Studies in East African Geography and Development. London: Heinemann, 259-267.

________, (ed.),1973. Studies in Famines and Food Shortages, special number, Journal of the Geographical Association of Tanzania, 8 (October).

________, 1975. Drought and the Optimization of Tanzania's Environmental Potential, paper prepared for Swedish Institute of Ecological Studies.

Mathewson, W. G., 1962. Monthly Report, Iringa District, Tanganyika, 26 February, typescript.

May, J. M., and D. L. McLellan, 1970. The Ecology of Malnutrition in Eastern Africa and Four Countries in Western Africa: Equatorial Guinea, the Gambia, Liberia, Sierra Leone, Malawi, Rhodesia, Zambia, Kenya, Tanzania, Uganda, Ethiopia, the French Territory of Afars and Issas, the Somali Republic and Sudan . New York: Hafner Publishing Co.

Mbithi, P. M., 1977. "Human factors in agricultural management in East Africa," Food Policy, 2 (February), 27-33.

Mbithi, P. M., and B. Wisner, 1973. "Drought and famine in Kenya," Journal of East African Research and Development, 3, 113-143.

McHenry, D. E., Jr., 1976. "The ujamaa village in Tanzania: A comparison with Chinese, Soviet and Mexican experiences in collectivization," Comparative Studies in Society and History, 18 (July), 347-370.

McLoughlin, P. F. M., 1967. Research on Agricultural Development in East Africa. New York: Agricultural Development Council.

Ministry of Finance and Planning, National Accounts of Tanzania, 1964-1972.

Monckeberg, F., 1977. "Impacts of food availability on health," Proceedings of the World Food Conference of 1976, June 27-July 1, Iowa State University, Ames, Iowa.

Monteith, J. L., 1965. "Light distribution and photosynthesis in field crops," Annals of Botany, N.S. 29, 17-37.

________, 1972. "Solar radiation and productivity in tropical ecosystems," Journal of Applied Ecology, 9, 747-766.

Mörth, H. T., 1967. Investigations into the Meteorological Aspects of the Variations in the Level of Lake Victoria, East African Meteorological Department Memoirs, Vol. 4.

Munro, J. F., 1975. Colonial Rule and the Kamba. Oxford: Clarendon Press.

Muthoka, D. K., 1978. Maize Improvement in Kenya. Kitale, Kenya: National Agricultural Research Station.

Myers, P. F., 1976. 1976 World Population Data Sheet. Washington, D. C.: Population Reference Bureau.

Myers, R. H., 1976, personal communication, 7 December.

Ndorobo, B., 1973. "Famine Problems in Dodoma District," Journal of the Geographical Association of Tanzania, 8 (October), 26-47.

Newman, J. L., 1973. "Hazards, adjustments, and innovation in central Tanzania," Rural Africana, 19, 4-19.

________, 1975. "Dimensions of Sandawe diet," Ecology of Food and Nutrition, 4, 33-39.

Nyerere, J. K., 1967a. The Arusha Declaration and TANU's Policy on Socialism and Self-Reliance. Dar es Salaam: The Government Printer.

________, 1967b. "Ujamaa--the basis of African socialism," in Freedom and Unity (Uhuru na Umoja). Dar es Salaam: Oxford University Press.

________, 1977. The Arusha Declaration Ten Years After. Dar es Salaam: The Government Printer.

Odingo, R. S., 1971. "Settlement and rural development in Kenya," in S. H. Ominde (ed.), Studies in East African Geography and Development. London: Heinemann, 162-176.

Omari, C. K., 1976. Strategy for Rural Development. Dar es Salaam: East African Literature Bureau.

Ominde, S. H., 1968. Land Population Movements in Kenya. London: Heinemann.

________, 1971. "The semi-arid and arid lands of Kenya," in S. H. Ominde (ed.), Studies in East African Geography and Development. London: Heinemann, 146-161.

Owen, D. F., 1973. Man in Tropical Africa. New York: Oxford University Press.

Palmer. B. C., et al., 1978. Analysis and Project Identification, Report No. 1, Kenya Marginal/Semi-arid Lands Pre-investment Inventory, Nairobi.

Patton, M., 1971. Dodoma Region, 1929-1959: A History of Famine, Research Report No. 44, Bureau of Resource Assessment and Land Use Planning, University of Dar es Salaam, Tanzania.

Penman, H. L., 1963. Vegetation and Hydrology, Technical Communication No. 53. Harpenden, U.K.: Commonwealth Bureau of Soils.

Polanyi, K., in G. Dalton (ed.), 1971. Primitive, Archaic and Modern Economies: Essays of Karl Polanyi. Boston: Beacon Press.

Porter, P. W., 1965. "Environmental potentials and economic opportunities - A background for cultural adaptation," American Anthropologist, Vol. 67, 409-420.

________, 1966. "East Africa - Population Distribution - As of August 1962," Annals of the Association of American Geographers, 56 (March), 180 and map supplement.

________, 1974. Potential Photosynthesis and Agriculture in Tanzania, Research Paper No. 29, Bureau of Resource Assessment and Land Use Planning, University of Dar es Salaam, Tanzania.

________, 1976. "Climate and agriculture in East Africa," in C. G. Knight and J. L. Newman (eds.), Contemporary Africa, Geography and Change. Englewood Cliffs, New Jersey: Prentice-Hall, Inc., 112-139.

Pratt, C., 1975. "Nyerere on the transition to socialism in Tanzania," The African Review, 5, 63-76.

Rald, J., 1970. Ujamaa: Problems of Implementation (Experiences from West Lake), Research Report No. 10, Bureau of Resource Assessment and Land Use Planning.

Rechcígl, M., Jr., 1975. World Food Problem: A Selective Bibliography of Reviews. Cleveland: CRC Press.

Rigby, P., 1969. Cattle and Kinship Among the Gogo. Ithaca: Cornell University Press.

Rusch, W. H., and T. H. Rose, 1977. Kenya: Smallholders Production Services and Credit Baseline Survey, Agricultural Year 1975-76, Executive Summary and Analysis. Report prepared for Government of Kenya and USAID by American Technical Assistance Corporation.

Ruthenberg, H. (ed.), 1968. Smallholder Farming and Smallholder Development in Tanzania, Afrika-Studien Nr. 24. München: Ifo-Institut für Wirtschaftsforschung.

Rutishauser, I. H. E., 1974. "Factors affecting the intake of energy and protein by Uganda preschool children," Ecology of Food and Nutrition, 3, 213-222.

Schlage, C., 1968. Polished versus Whole Maize: Some Nutritional and Economic Implications of the Traditional Processing of Maize in Northeastern Tanzania. Research Paper No. 2, Bureau of Resource Assessment and Land Use Planning, University of Dar es Salaam.

Scientific American, 1976. Food and Agriculture. Special issue, 235.

Scrimshaw, N. S., 1966. "Food: World Problems," in David L. Sils (ed.), International Encyclopaedia of the Social Sciences, 5. New York: Macmillan, 502-507.

Scrimshaw, N. S., and M. Béhar, 1974. Nutrition and Agricultural Development. New York: Plenum Press.

Shahid, J. B., and T. J. Goering, 1977. "Food Problems of the Low-Income Countries," Finance and Development, Vol. 14 (June), 15-18.

Sharman, A., 1970. "Nutrition and social planning." The Journal of Development Studies, 6 (July), 76-91.

Shivji, I. G., et al., 1973. The Silent Class Struggle. Dar es Salaam: Tanzania Publishing House.

Shivji, I. G., 1976. Class Struggles in Tanzania. New York: Monthly Review Press.

Simoons, F. J., 1974. "Rejection of fish as human food in Africa - A problem in history and ecology." Ecology of Food and Nutrition, 3, 89-105.

Southworth, H. M., and B. F. Johnston, 1967. Agricultural Development and Economic Growth. Ithaca: Cornell University Press.

Swynnerton, R. J. M., 1955. A Plan to Intensify the Development of African Agriculture in Kenya. Nairobi: The Government Printer.

Tallantire, A. C., 1975. "A preliminary study of the food plants of the West Nile and Madi Districts of Uganda," East African Agricultural and Forestry Journal, 40, 233-255.

Taylor, D. R. F., 1974. "Land reform in Kenya: A reappraisal," Rural Africana, 23 (Winter), 79-90.

The Standard, 26 June 1971.

Thomas, I. D., 1972. Infant Mortality in Tanzania, Department of Geography, University of Dar es Salaam.

Thomson, J., 1885. Through Masai Land: A Journal of Exploration Among the Snowclad Volcanic Mountains and Strange Tribes of Eastern Equatorial Africa. London: Sampson Low, Marston, Searle, and Rivington.

Turnbull, C. M., 1972. The Mountain People. New York: Simon and Schuster.

Uganda Protectorate, 1961. Estimation of Cultivated Areas, 1960, Statistics Branch, Ministry of Economic Development, December.

United Republic of Tanzania, 1967. Preliminary Results of the Population Census Taken on the Night of 26-27th, August, 1967, Central Statistical Bureau, Ministry of Economic Affairs and Development Planning, Dar es Salaam.

United Republic of Tanzania, 1969. 1967 Population Census, Vol. 1, Statistics for Enumeration Areas, Central Statistical Bureau, 1969.

Van de Laar, A., 1972. "Foreign business and capital export from developing countries: The Tanzanian experience," in L. Cliffe and J. S. Saul (eds.), Socialism in Tanzania, Vol. 1, Politics, Dar es Salaam: East African Publishing House, 83-87.

Wang'ati, F. J., 1969. "Methods of estimating photosynthesis in the field and their application in land use planning." Proceedings, 4th Specialist Meeting on Applied Meteorology, East African Agriculture and Forestry Research Organization.

West, Q. M., 1974. "Food shortages: A poor man's energy crisis," Ecology of Food and Nutrition, 3, 243-249.

White, G. F. (ed.), 1974. Natural Hazards: Local, National and Global. New York: Oxford University Press.

Willett, J. W., 1976. The World Food Situation: Problems and Prospects to 1985. Dobbs Ferry, N. Y.: Oceana Publications.

Winans, E. V., 1977. Personal communication, 31 March.

Winans, E. V., 1978. Personal communication, 15 February.

Wisner, B., and P. M. Mbithi, 1972. Drought in Eastern Kenya: Comparative Observations of Nutritional Status and Farmer Activity at 17 Sites. Discussion Paper No. 167, Institute of Development Studies, University of Nairobi.

________, 1974. "Drought in Eastern Kenya: Nutritional Status and Farmer Activity," in Gilbert F. White (ed.), Natural Hazards. New York: Oxford University Press, 87-97.

Wisner, B., 1976. An Overview of Drought in Kenya. Natural Hazard Research Working Paper 30.

________, 1977a. The Human Ecology of Drought in Eastern Kenya. Unpublished Ph.D. dissertation, Clark University.

________, 1977b. Personal communication, 6 September.

Wolff, R. D., 1974. The Economics of Colonialism - Kenya and Britain, 1870-1930. New Haven: Yale University Press.

Woodhead, T., 1968a. Studies of Potential Evaporation in Kenya. Republic of Kenya, Ministry of Natural Resources, Water Development Department.

________, 1968b. Studies of Potential Evaporation in Tanzania. East African Common Services Organization in Cooperation with the Water Development and Irrigation Division, Ministry of Lands, Settlement and Water Development, Tanzania.

World Food Conference of 1976, 1977. Proceedings, June 27-July 1, Iowa State University, Ames, Iowa.

Young, R., and H. Fosbrooke, 1960. Smoke in the Hills. Evanston: Northwestern University Press.

FOREIGN AND COMPARATIVE STUDIES PROGRAM
Maxwell School of Citizenship and Public Affairs
Syracuse University
119 College Place
Syracuse, NY 13210

EASTERN AFRICAN SERIES

No.	Title	Author(s)	Price
II.	The Conflict Over What Is To Be Learned in Schools: A History of Curriculum Politics in Africa 1971 113 pp.	Stephen P. Heyneman	$ 4.50
V.	The Politics of Indifference : Portugal and Africa, A Case Study in American Foreign Policy 1972 41 pp.	John A. Marcum	2.50
VIII.	The Zimbabwe Controversy: A Case of Colonial Historiography 1973 142 pp.	David Chanaiwa	4.50
IX.	The Union of Tanganyika and Zanzibar: A Study in Political Integration 1973 114 pp.	Martin Bailey	4.50
X.	The Pattern of African Decolonization: A New Interpretation 1973 123 pp.	Warren Weinstein John J. Grotpeter	4.50
XII.	Protest Movements in Colonial East Africa: Aspects of Early African Response to European Rule 1973 96 pp. & xi.	Robert Strayer Edward I. Steinhart Robert Maxon	4.50
XIII.	Education for What? British Policy Versus Local Initiative 1973 100.pp. & viii.	Charles H. Lyons Kenneth J. King Richard D. Heyman John M. Trainor	4.50
XIV.	Ethnicity in Contemporary Africa 1973 59 pp. & vi.	Robert H. Bates	3.50
XVI.	Africans in European Eyes: The Portrayal of Black Africans in Fourteenth and Fifteenth Century Europe 1975 98 pp.	Peter A. Mark	4.50
XVII.	Drought, Famine and Population Movements in Africa 1975 144 pp. & vi.	James L. Newman, ed.	4.50
XVIII.	Technology for Ujamaa Village Development in Tanzania 1975 64 pp.	David J. Vail	3.50
XIX.	Citizenship in Africa: The Role of Adult Education in the Political Socialization of Tanganyikans, 1891-1961 1975 119 pp.	Joel C. Millonzi	4.50
XX.	Health Care Financing in Mainland Tanzania 1975 104 pp.	Manuel Gottlieb	4.50
XXI.	Three Aspects of Crisis in Colonial Kenya 1975 91 pp. & xxiii.	Bismarck Myrick David L. Easterbrook Jack Roelker	4.50
XXII.	Africa and International Crises 1976 106 pp.	Robert Brown Donald F. Heisel Charles H. Lyons Harvey Flad	4.50
XXIV.	Political Identity: A Case Study from Uganda 1976 179 pp.	Marshall H. Segall Martin Doornbos Clive M. Davis	5.50

XXV.	East African Culture History 1976 93 pp.	Joseph Gallagher,ed. Iris Berger Carole A. Muchanan Thomas J. Hinnebusch	4.50

AFRICAN SERIES

XXVI.	Linguistic Diversity and Language Belief in Kenya: The Special Position of Swahili 1977 127 pp.	John Rhoades	6.00
XXVII.	Explorations in Quantitative African History 1977 190 pp.	Joseph Smaldone,ed. Bruce Fetter James D. Graham David Henige Robert A. Hess Patrick Manning William Schneider	7.00
XXVIII.	Wildlife in Tanzanian Settlement Policy: The Case of the Selous 1977 122 pp.	Gordon Matzke	6.50
XXIX.	On His Majesty's Service: The Origin of Uganda's African Civil Service, 1912-1940 1977 72 pp.	Nizar Motani	5.00
XXX.	A Comparative Study of Political Involvement in Three African States: Botswana, Ghana and Kenya Summer 1978 139 pp. & ix.	John D. Holm Fred M. Hayward Mark W. Graesser Marc Howard Ross	6.50
XXXI.	Afocha: A Link Between Community and Administration in Harar, Ethiopia Late Summer 1978	Peter Koehn Sidney R. Waldron	7.00
XXXII.	Food and Development in the Semi-Arid Zone of East Africa 1979 110 pp.	Philip W. Porter	7.00
XXXIII.	Islamization Among the Upper Pokomo, 2nd Edition 1980 128 pp.	Robert L. Bunger,Jr.	8.00
XXXIV.	Kiziba: The Cultural Heritage of an Old African Kingdom 1980 104 pp.	Abel G.M. Ishumi	7.00
XXXV.	Traditional Health Care Delivery in Contemporary Africa	Priscilla Ulin and Marshall Segall, eds. Charles M. Good Herbert Rappaport Anita Spring	tba

SPECIAL PUBLICATIONS

	Title	Author(s)	Price
1.	Basic Structure of Swahili 1966 151 pp.	James L. Brain	4.50
2.	Basic Structure of Swahili, Part II 1969 34 pp.	James L. Brain	2.50
3.	A Social Science Vocabulary of Swahili 1968 43 pp.	James L. Brain	2.50
4.	A Short Dictionary of Social Science Terms for Swahili Speakers 1969 70 pp.	James L. Brain	3.50
5.	Modern Makonde Sculpture Exhibit Catalog 1968 103 pp.		3.50
6.	Shindano: Swahili Essays and Other Stories 1971 55 pp.	Alice Grant Johannes C. Mlela Jean F. O'Barr William O'Barr	2.50
7.	African Microfilms at the E.S. Bird Library, Syracuse University: An Annotated Guide 1975 72 pp.	David L. Easterbrook Kenneth P. Lohrentz	3.50

EASTERN AFRICAN BIBLIOGRAPHIC SERIES

	Title	Author(s)	Price
1.	A Bibliography of Malawi 1965 161 pp.	Edward F. Brown Carol A. Fisher John B. Webster	5.00
2.	A Bibliography on Kenya 1967 461 pp.	Shirin G. Kassam Robert S. Peckham Barbara A. Skapa John B. Webster	8.00
3.	The Guide to the Kenya National Archives 1969 452 pp.	Robert G. Gregory Robert Maxon Leon Spencer	16.00

OCCASIONAL BIBLIOGRAPHIES

	Title	Author(s)	Price
18.	A Guide to the Coast Province Microfilm Collection Kenya National Archives Kenya Seyidie (Coast) Province, Correspondence and Reports, 1891-1962 1971 191 pp.	Harvey Soff	6.00
19.	Microfilms Related to Eastern Africa, Part I (Kenya). A Guide to Recent Acquisitions of Syracuse University 1971 88 pp.	R.F. Morton Harvey Soff	4.50
21.	Microfilms Related to Eastern Africa, Part II (Kenya, Asian and Miscellaneous) A Guide to Recent Acquisitions of Syracuse University 1973 142 pp.	David Leigh R.F. Morton	7.00
22.	A Guide to the Nyanza Province Microfilm Collection, Kenya National Archives, Part I: Section 10B, Correspondence and Reports, 1925-1960 1974 137 pp.	Alan C. Solomon	5.00

No.	Title	Author(s)	Price
23.	A Select Bibliography of Soviet Publications on Africa in General and Eastern Africa in Particular, 1972-73 1974 43 pp. Supplement V	Ladislav Venys	2.50
24.	A Guide to the Nyanza Province Microfilm Collection, Kenya National Archives, Part II Section 10A Correspondence and Reports, 1899-1942 1974 137 pp.	Alan C. Solomon	3.50
25.	A Guide to Nyanza Province Microfilm Collections, Kenya National Archives, Part III: Section 10, Daily Correspondence and Reports, 1930-1963, Vol. 1 1975 258 pp. & vi.	Kenneth P. Lohrentz Alan C. Solomon	8.00
26.	A Guide to Nyanza Province Microfilm Collection, Kenya National Archives, Part III: Section 10, Daily Correspondence and Reports, 1930-1963, Vol. II 1975 254 pp. & vi.	Kenneth P. Lohrentz Alan C. Solomon	8.00
27.	Microfilms Relating to Eastern Africa, Part III (Kenya and Miscellaneous): A Guide to Recent Acquisitions of Syracuse University 1975 112 pp. & iv.	David L. Easterbrook Alan C. Solomon Thomas F. Taylor	7.00

Publications may be orderd from: Publications Desk
Foreign and Comparative Studies Program
119 College Place
Syracuse, NY 13210